To Bill and Carleen Glasser, the "first couple"
of choice theory and reality therapy

And also to our spouses
Rich Robey, Sandie Wubbolding, and Laura Carlson
With much gratitude for your loving support, patience, and
unfaltering commitment to our marriage relationships

This book is part of the Family Therapy and Counseling Series, edited by Jon Carlson.

Routledge
Taylor & Francis Group
711 Third Avenue
New York, NY 10017

Routledge
Taylor & Francis Group
27 Church Road
Hove, East Sussex BN3 2FA

© 2012 by Taylor & Francis Group, LLC
Routledge is an imprint of Taylor & Francis Group, an Informa business

Printed in the United States of America on acid-free paper
Version Date: 20120106

International Standard Book Number: 978-0-415-89125-7 (Hardback)

For permission to photocopy or use material electronically from this work, please access www.copyright.com (http://www.copyright.com/) or contact the Copyright Clearance Center, Inc. (CCC), 222 Rosewood Drive, Danvers, MA 01923, 978-750-8400. CCC is a not-for-profit organization that provides licenses and registration for a variety of users. For organizations that have been granted a photocopy license by the CCC, a separate system of payment has been arranged.

Trademark Notice: Product or corporate names may be trademarks or registered trademarks, and are used only for identification and explanation without intent to infringe.

Library of Congress Cataloging-in-Publication Data

Contemporary issues in couples counseling : a choice theory and reality therapy
  approach / [edited by] Patricia A. Robey, Robert E. Wubbolding, Jon Carlson.
    p. cm. -- (Family therapy and counseling)
  Summary: "Contemporary Issues in Couples Counseling addresses the most
common and difficult issues that people in the helping professions face when using
CBT with couples--and provides concrete solutions for addressing them effectively. In
it, clinicians will find a handy reference for professionals who are looking for useful
information and skills that can be applied immediately in their sessions. The book
uses the time-tested, evidence-based strategies for helping clients focus on the here
and now, not the past, and for helping clinicians create effective treatment plans and
ensure that that clients meet their individual needs while also addressing the needs of
their partners"-- Provided by publisher.
  Includes bibliographical references and index.
  ISBN 978-0-415-89125-7 (hardback)
  1. Couples--Counseling of. 2. Reality therapy. 3. Control theory. I. Robey, Patricia
A. II. Wubbolding, Robert E. III. Carlson, Jon.

BF636.7.G76C66 2012
616.89'1562--dc23   2011032267

Visit the Taylor & Francis Web site at
http://www.taylorandfrancis.com

and the Routledge Web site at
http://www.routledgementalhealth.com

Contemporary Issues in Couples Counseling

A Choice Theory and Reality Therapy Approach

Edited by
Patricia A. Robey,
Robert E. Wubbolding,
and Jon Carlson

 Routledge
Taylor & Francis Group
New York   London

Contemporary Issues
in Couples Counseling

The Family Therapy and Counseling Series
Series Editor
Jon Carlson, Psy.D., Ed.D.

Kit S. Ng
Global Perspectives in Family Therapy: Development, Practice, Trends

Phyllis Erdman and Tom Caffery
Attachment and Family Systems: Conceptual, Empirical, and Therapeutic Relatedness

Wes Crenshaw
Treating Families and Children in the Child Protective System

Len Sperry
Assessment of Couples and Families: Contemporary and Cutting-Edge Strategies

Robert L. Smith and R. Esteban Montilla
Counseling and Family Therapy With Latino Populations: Strategies That Work

Catherine Ford Sori
Engaging Children in Family Therapy: Creative Approaches to Integrating Theory and Research in Clinical Practice

Paul R. Peluso
Infidelity: A Practitioner's Guide to Working With Couples in Crisis

Jill D. Onedera
The Role of Religion in Marriage and Family Counseling

Christine Kerr, Janice Hoshino, Judith Sutherland, Sharyl Parashak, and Linda McCarley
Family Art Therapy

Debra D. Castaldo
Divorced Without Children: Solution Focused Therapy With Women at Midlife

Phyllis Erdman and Kok-Mun Ng
Attachment: Expanding the Cultural Connections

Jon Carlson and Len Sperry
Recovering Intimacy in Love Relationships: A Clinician's Guide

Adam Zagelbaum and Jon Carlson
Working With Immigrant Families: A Practical Guide for Counselors

David K. Carson and Montserrat Casado-Kehoe
Case Studies in Couples Therapy: Theory-Based Approaches

Shea M. Dunham, Shannon B. Dermer, and Jon Carlson
Poisonous Parenting: Toxic Relationships Between Parents and Their Adult Children

Bret A. Moore
Handbook of Counseling Military Couples

Len Sperry
Family Assessment: Contemporary and Cutting-Edge Strategies, 2nd ed.

Patricia A. Robey, Robert E. Wubbolding, and Jon Carlson
Contemporary Issues in Couples Counseling: A Choice Theory and Reality Therapy

CONTENTS

SERIES EDITOR'S FOREWORD

> What happened in the past that was painful has a great deal to do with what we are today, but revisiting this painful past can contribute little or nothing to what we need to do now.

William Glasser (1998, p. 334)

In 1969, my wife Laura and I visited Los Angeles following the APGA (now ACA) Convention that was held in Las Vegas. We were recently married and didn't have much money (especially after one week in Las Vegas!). We had prepaid our LA trip and were finding that money was going to be tight on this trip extension. We decided to walk around and see the sights near the hotel as that was free. We came upon a local bookstore with one of the two front-display windows full of *Reality Therapy* and *Schools Without Failure,* both by William Glasser, MD. I was mesmerized, as I had heard about Dr. Glasser being one of the top therapists but knew nothing about his work. Laura must have seen the look on my face and she said (knowing we couldn't afford it) that I should get one. I chose *Reality Therapy* and spent all of my free time on the trip reading, although, as Laura reminded me, we had to take public transportation everywhere as the cost of the book had impacted our meager sightseeing budget.

Since that time I have read all of Dr. Glasser's books and become friends with him and his wife, Carleen. I have marveled at the brilliant simplicity of his ideas and have found myself frequently using his ideas to explain behavior or to create an intervention with individuals and couples. Most professionals see Glasser's ideas as having value for individuals as well as for school guidance but have been unaware of how effective they can be with couples.

This book addresses that knowledge gap and shows how effective reality therapy and choice theory can be with couples. I hope that the readers are open to these ideas and appreciate their application with a complete range of the issues that face today's couples. It is only fitting that the lead editors, Pat Robey and Bob Wubbolding, have trained therapists around the world in these important ideas. The chapters have been carefully selected, and a group of exceptional professionals share their insights as to how to work effectively with couples from this viewpoint. As William Glasser asserts, "We almost always have choices, and the better the choice, the more we will be in control of our lives" (1998).

Jon Carlson, PsyD, EdD, ABPP
Distinguished Professor
Governors State University

REFERENCE

Glasser, W. (1998). *Choice theory*. New York: HarperCollins.

PREFACE

Everyone agrees: Times are difficult and we stand at the fork in the road. We can consume our time with worrying, fretting, complaining, and rebelling against what everyone agrees is an uncertain future. Another option is to focus attention on what we can control. From the perspective of choice theory and reality therapy, human beings can control only their own behaviors. If athletes control their own choices during the game and pay less attention to the score or the apparent outcome, their performance is more effective than if they focus on the future results of their toil. If a baseball batter focuses his attention on the outfield fence and not on the incoming pitch, he will likely strike out. Similarly, if a soccer player looks to the goal rather than the ball, the kick is less likely to result in a score.

It is easy for a couple continuously to discuss topics over which they have little control. These might include job security, possible unemployment, erosion of savings, foreclosures, future education of children, family loss, sickness, and a host of economic worries. This book describes the alternative path at the fork in the road with the hope that it will be *the road taken*. This path is a road paved with a sense of hope and joy, a rewarding journey but not without detours, delays, and potholes. Yet there are rest stops that provide tools for coping with unavoidable uncertainties, contingencies, obstacles, and emergencies. Couples can control how they spend their time together, the topics of conversation, and their attitudes toward the world around them, as well as how they relate to each other.

This book presents a wide range of applications of choice theory and reality therapy. The book also includes an explanation of choice theory and reality therapy as a means for gaining practical ideas immediately usable in human relationships. Part I begins with an overview of choice theory and reality therapy, including the application of the WDEP system to counseling. This chapter provides a context for understanding specific applications offered in subsequent chapters. Also in this section is an interview with William and Carleen Glasser, who discuss how they apply choice theory and reality therapy in their own relationship. Dr. Glasser first formulated reality therapy in a mental hospital and in a correctional institution in the 1950s and 1960s. Since the early days of its application to severe mental health and behavioral issues, it has been applied to counseling, education, management, and virtually every human relationship.

Part II of the book emphasizes the issues couples face and, through hypothetical or compilation case studies, provides practical strategies and techniques that counselors can apply in their work. The authors are experts in counseling, counselor education, and choice theory and reality therapy. Collectively, they have used reality therapy and choice theory successfully with thousands of clients. The topics in this section address a broad range of issues. One of these issues is addiction and recovery. The author explains how reality therapy complements the 12 steps of Alcoholics Anonymous and Al-Anon. Several chapters focus on issues related to diversity, including challenges facing interfaith couples, multicultural couples, aging couples, gay couples, and military couples. Other chapters focus on mental health issues, chronic illness, parenting conflicts, sexuality, infidelity, medication management, and power conflicts.

Part III focuses on proactive approaches to building solid couple relationships. Counselors will learn how to facilitate workshops in which couples learn how to bring out the best in one another before they enter committed relationships. Using reality therapy along with traditional tools like individual genograms is another proactive technique that can help couples define their relationships. The author describes how assessing the pictures of what is important to them can help couples get their relationships off to a good start. The third chapter in this section demonstrates how couples can be taught the skill of negotiating to create common pictures where differences exist. This exercise teaches couples to communicate their specific wants and learn how to compromise and negotiate.

Part IV summarizes the overall map of more effective living provided in Parts I through III and presents a vision of the future.

Contemporary Issues in Couples Counseling: A Choice Theory and Reality Therapy Approach demonstrates that the ideas formulated by William Glasser are applicable in many ways to individuals seeking to fulfill their need for belonging through intimate relationships. Participants attending training seminars often ask how reality therapy connects with family life. This book answers this question and points the reader toward the far reaching frontiers and the systemic applicability of reality therapy as used by practitioners trained in choice theory and reality therapy.

Robert E. Wubbolding, EdD

EDITORS

Patricia A. Robey, EdD, LPC, NCC, CTRTC, is an assistant professor of counseling and program coordinator for the Masters in Counseling Program Coordinator at Governors State University, University Park, Illinois. She is a licensed professional counselor and specializes in applying reality therapy and choice theory in her work with individuals, couples, families, and groups. Dr. Robey is a senior faculty member of the William Glasser Institute and has taught the concepts of choice theory and reality therapy in the United States and internationally.

Robert E. Wubbolding, EdD, is a psychologist, a professional clinical counselor, and a member of the American Psychological Association. He is also professor emeritus of counseling at Xavier University and director of the Center for Reality Therapy in Cincinnati, Ohio. Dr. Wubbolding served as the director of training and chair of the William Glasser Institute Professional Development Committee 1988–2011. He has written over 140 articles, essays, and chapters in textbooks as well as 11 books and has published 12 videos on reality therapy. His widely acclaimed books include *Reality Therapy for the 21st Century, A Set of Directions for Putting and Keeping Yourself Together,* and *Reality Therapy: Theories of Psychotherapies Series* (published by the American Psychological Association).

Jon Carlson, PsyD, EdD, ABPP, is distinguished professor of psychology and counseling at Governors State University, University Park, Illinois, and a psychologist with the Wellness Clinic in Lake Geneva, Wisconsin. Jon has created over 300 professional videos used in universities around the world for the training of counselors, therapists, and psychologists. He has authored over 55 books in the areas of family therapy, marital enrichment, consultation, and Adlerian psychology. Some of his best known works include *Recovering Intimacy, Adlerian Therapy, Never Be Lonely Again,* and *Time for a Better Marriage.* He has received many professional awards, including being named a "living legend" by the American Counseling Association and the Lifetime Achievement Award for Education and Training from the American Psychological Association.

CONTRIBUTORS

Sylinda Gilchrist Banks, EdD, is an associate professor at Norfolk State University in Norfolk, Virginia. Dr. Banks is certified in reality therapy, a Glasser scholar, and a faculty member of the William Glasser Institute. She has coauthored several books and authored *Using Choice Theory and Reality Therapy to Enhance Student Achievement* (2009). Dr. Banks's research interests include school counseling program development and accountability, self-care, and the application of choice theory/reality therapy in educational settings.

Mark J. Britzman, EdD, LP, NCC, CCMHC, CTRTC, is a tenured professor in the Counseling and Human Resource Department at South Dakota State University. Dr. Britzman is a Glasser Scholar and a national trainer for the Josephson Institute of Ethics; and founder of the CHARACTER COUNTS! program located in Los Angeles, California. He is also is the owner of Pursuing the Good Life: Professional Counseling and Consultation Services located in Sioux Falls, South Dakota.

Nancy S. Buck, PhD, is a developmental psychologist and a senior faculty member of the William Glasser Institute. As an expert in relationships with an emphasis on the parent/child relationship, she founded Peaceful Parenting Inc. She has authored two groundbreaking parenting books: *Peaceful Parenting* and *Why Do Kids Act That Way? The Instruction Manual Parents Need to Understand Children at Every Age.* Dr. Buck is an international speaker and trainer and a blogger for *Psychology Today.*

Thomas K. Burdenski, Jr., PhD, LPC, LMFT, CTRTC, is an assistant professor at Tarleton State University, Stephenville, Texas, where he teaches graduate courses in counseling and psychology, including marriage and family therapy, addictions counseling, and brief therapy. Dr. Burdenski is licensed as a marriage and family therapist, a professional counselor, and a psychologist. He is a faculty member and a U.S. Advisory Board member of the William Glasser Institute and serves on the editorial board of the *International Journal of Choice Theory and Reality Therapy.*

Willa J. Casstevens, PhD, MSW, is an assistant professor at the North Carolina State University Department of Social Work in Raleigh, North Carolina. She received her doctorate in social welfare from the School of Social Work at Florida International University. Her research involves psychosocial treatments and consumer advocacy in mental health, and she recently published her first book, *A Mentored Self-Help Intervention for Psychotic Symptom Management.* Dr. Casstevens is a licensed clinical social worker in Florida and North Carolina and certified in choice theory/reality therapy by the William Glasser Institute.

Gloria Smith Cissé, MSW, is a licensed master social worker and licensed professional counselor, is certified in reality therapy, and is a Glasser scholar and faculty member of the William Glasser Institute. Ms. Cissé is currently a private practitioner with the Southern Center for Choice Theory in Georgia. Her research interests include prevention and treatment of interpersonal violence, the application of choice theory/reality therapy in clinical/community settings, and supervision.

Jeri L. Crowell, EdD, is a licensed professional counselor in Georgia, is certified in reality therapy, and is a Glasser scholar and faculty member of the William Glasser Institute. Dr. Crowell is currently a core faculty member at Capella University, Minneapolis, Minnesota, in mental health counseling. Research interests include group work (coauthor of *Group Techniques: How to Use Them More Purposefully,* 2008), crisis intervention, and the application of reality therapy/choice theory in counseling settings.

Jill D. Duba, PhD, is a licensed professional clinical counselor, certified in reality therapy, trained in EMDR, and an associate professor in the Department of Counseling at Student Affairs at Western Kentucky University in Bowling Green. She also maintains a small private practice. Jill has published various book chapters and articles related to counseling religious couples and interfaith couples, including *Therapy With Religious Couples; The Basic Needs Genogram: A Tool to Help Inter-religious Couples Negotiate;* and *Treating Infidelity: Considering Narratives of Attachment.* Book chapters include "Intimacy and the Recovery of Intimacy in Religious and Faith-based Relationships" in *Recovering Intimacy in Love Relationships: A Clinician's Guide* and "Inter-religion Marriages" in *The Role of Religion in Marriage and Family Counseling.*

Terri Earl-Kulkosky, PhD, is an associate professor of social work at Fort Valley State University in Georgia. She received her master's in social work and PhD in child and family development/marriage and family therapy from the University of Georgia. Her research efforts have focused on African American women and the impact of cultural, historical, and societal factors on coping strategies, health dispari-ties, and family roles. She has also explored the ethical dilemmas that rural and minority therapists face dealing with dual relationships and family therapy interventions. She has taught and developed classes on cultural diversity, rural human services, family violence intervention, gender studies, and social work practice methods. Dr. Earl-Kulkosky worked in community mental health, hospice, and private practice before working full time in academia. Her practice experience included pioneering work on child abuse protocols for interviewing children that were victims of abuse.

Janet Fain-Morgan, EdD, NCC, LPC, CTRTC, is in private practice in Columbus, Georgia, primarily counseling active duty military sol-diers and family members. As well as enjoying private practice with Columbus Psychology Associates, Janet has served as the chairwoman of the Muscogee County Domestic Violence Roundtable Education and Outreach Committee and occupied a position on the executive board. Janet's relationship with the military includes serving as a medic in the U.S. Army Reserves, being the daughter of a servicemem-ber, and now, the mother of a servicemember (with two deployments to a combat zone). Janet is a member of the William Glasser Institute (WGI), the National Catholic Ministry to the Bereaved, the Licensed Professional Counselors Association of Georgia (LPCGA), the National Board of Certified Counselors (NBCC), and the American Counseling Association (ACA).

Maureen Craig McIntosh, a former registered nurse, has more than 20 years of experience in the field of sexual health for the province of New Brunswick, Canada. She is a Canadian certified counselor and a certified personal and executive coach, as well as a licensed emotional fitness coach. She is a member of the senior faculty of the William Glasser Institute and has a private counseling, training, and coaching practice. She owns her own company, Moncton Reality Therapy Consultants.

Neresa B. Minatrea, PhD, is a licensed professional clinical counselor and a certified addictions counselor. She is also certified in reality therapy and is a professor in the Department of Counseling at Student Affairs at Western Kentucky University in Bowling Green. Dr. Minatrea also maintains a private practice and has published various articles and book chapters. She currently serves on the Kentucky Professional Counseling Licensing Board.

Jerry A. Mobley, PhD, is a licensed professional counselor trained in marriage and family therapy and is currently the interim chair of the School Counselor Education program at Fort Valley State University in Georgia. Dr. Mobley's research interests include theoretical training of counselors, group work, and addiction studies. He is the author of *An Integrated Existential Approach to Counseling Theory and Practice*.

Sela E. Nagelhout, MS, NCC, is a former registered nurse with several years of experience in critical care settings. She is a certified *Within Our Reach* instructor and is in private practice at Pursuing the Good Life Professional Counseling and Consultation Services, Sioux Falls, South Dakota.

Kim Olver, MS, LCPC, NCC, is a senior faculty member of and the executive director for the William Glasser Institute. She has trained thousands of social workers, counselors, and therapists in Glasser's ideas and concepts. Kim is the founder of Coaching for Excellence, a life coach and the award-winning, best-selling author of *Secrets of Happy Couples*. She has appeared in *Women's World*, *Whole Living*, *Counseling Today*, and *Fitness Magazine* as well as various radio shows and podcasts.

Brandi Roth, PhD, is a licensed psychologist in private practice in Beverly Hills, California. Dr. Roth specializes in conflict resolution with couples, adults, and families; in children's academic and behavioral challenges; and in comprehensive neuropsychological evaluations. Dr. Roth is also a William Glasser Institute faculty member. Dr. Roth previously worked in education as a classroom teacher, special education teacher, educational therapist, administrative specialist, and in-service instructor. She is an international consultant and speaker. Dr. Roth is the coauthor of four books. *Role Play Handbook: Understanding and Teaching the New Reality Therapy* teaches choice theory through role play. *Relationship Counseling With Choice Theory Strategies* presents strategies for couples counseling. *Choosing the Right School for Your Child, Revised 2008,* is a nationwide guide and workbook for families selecting elementary and secondary schools. *Secrets to School Success, Guiding Your Child Through a Joyous Learning Experience* provides practical tools to encourage children and families to have an effective and joyous school experience. Further information is available on her website at www.associationofideas.com.

Tammy F. Shaffer is a contributing faculty member at Walden University, where she teaches graduate courses in counseling, including internship, group counseling, theories, and marriage and family counseling. She is a licensed professional counselor in both Texas and Kentucky. She serves on the editorial board of the *Family Journal* and is a disaster mental health counselor volunteer with the American Red Cross; deployments have included Hurricane Katrina, and tornadoes that hit Smithville, Mississippi, in 2011. Research interests include sibling loss, trauma and disaster response, and music in reality therapy.

Vanessa L. White, MS, CTRTC, LPC, is a high-risk care manager for a behavioral health managed-care organization and serves as an adjunct faculty member at Marywood University in Scranton, Pennsylvania. Ms. White conducts trainings on the topics of diversity and the needs of adolescents, as well as the specific needs of gay, lesbian, bisexual, and transgender youths in the child welfare system. Ms. White is openly lesbian and lives with her daughter in northeast Pennsylvania.

I

Choice Theory and Reality Therapy

1

INTRODUCTION TO CHOICE THEORY AND REALITY THERAPY

Robert E. Wubbolding and Patricia A. Robey

INTRODUCTION

Why is there occasional or chronic pain in friendships, marriage, and all human relationships? How can individuals improve these relationships and increase intimacy, thereby diminishing stress? Choice theory and reality therapy offer both an explanation and possible choices for improving human alliances, interpersonal bonds for fulfilling the innate need for belonging. The use of choice theory and reality therapy provides both a map and specific useful tools for enhancing relationships, especially if both parties subscribe to and practice the principles contained in these life-changing ideas. A relationship built on choice theory and reality therapy leads to a sense of belonging, inner control, freedom, enjoyment, and increased contentment.

Relationships, especially marriage, fail when one of the partners attempts to control the other person to a degree unacceptable to that person. For example, one spouse dominates the other by refusing to negotiate differences and bulldozes decisions that affect both of them. When the submissive person believes that such controlling behaviors are unacceptable, the relationship begins to crumble. The second reason relationships fail is varying degrees of incompatibility in their respective quality worlds. For instance, a person with a high desire or want for freedom or independence might seek a lifestyle filled with autonomous activities or other behaviors from which the partner recoils and thus the relationship is strained, damaged, or even ruined.

CHOICE THEORY

Theologians might explain dysfunctional relationships as due to original sin. Because of human weakness, people are prone to shatter their lives to a greater or lesser degree. Sociologists would elaborate on the failing of society to provide proper supports and the temporary nature of relationships as evidenced in the mobility of families. Psychoanalytic therapists might emphasize the lack of balance between ego, id, and superego as well as unconscious conflicts due to early childhood relationships. Cognitive theory would emphasize irrational thinking as the root cause of conflict when partners indulge in self-talk statements such as *the world must arrange itself so that I experience only pleasure.*

Conversely, choice theory offers a positive but not naïve view of human nature—a comprehensive explanation of personal problems, personal growth, and interpersonal relationships. It stresses current motivation rather than past experiences. It emphasizes conscious drives rather than attempts to resolve unconscious and unresolved illusive fantasies. It is a "here and now" theory expressed in jargon-free language understandable to clients, students, and licensed professionals. People wishing to learn choice theory read books about it, attend training seminars, and seek professional consultation.

Origin

William Glasser, MD (1998), developed choice theory from a cybernetic theory that has existed for many decades. Norbert Wiener (1948, 1950) described the human mind as a negative feedback input control system. A rocket is such a system. When it "perceives" that it is heading off target, it provides information to the targeting mechanism that corrects its course. Similarly, human behavior aims at impacting the external world so as to gain input.

Glasser's immediate predecessor, William Powers (1973), expanded what was called "control theory" or "control system theory" in his landmark book, *Behavior, the Control of Perception.* He asserted that human behaviors originate *within* human beings and are therefore not thrust upon them from the external world. Furthermore, how human beings behave toward their external environment (i.e., the collection of their experiences toward their external world) determines their perceptions of it. Because of the emphasis on behavior as a choice and because of the introduction of five human needs driving human choice, Glasser (1998) changed the name of his interpretation of control theory to choice theory.

Human Motivation

As seen from the perspective of choice theory, the human mind functions as a control system. The analogy of the rocket helps to understand the goal centeredness of human behavior and its ability to correct itself. Another useful analogy is that of the thermostat. This mechanism controls its environment (i.e., gains a "perception" of its impact on the world around it). If the thermostat is set at 70 degrees, it sends a signal to the furnace or air conditioner to generate goal-centered and specific behaviors. The result is that the environment matches the thermostat's "desire." By analogy, it can be said that the thermostat has attained its goal of maneuvering its external world to make it congruent with its internal "wants." The thermostat, therefore, perceives that it is successful.

An interesting sidelight is that the control system can be fooled. If a match is held under the thermometer, the thermostat will read an inaccurate message and will mistakenly adjust the room temperature. A human relationship suffers stress and strain when one person functions under the erroneous illusion that "all is well" between the parties. Drugs, alcohol, gambling, and other addictions, as well as extramarital affairs, can appear to satisfy an individual and yet poison the relationship. It is as though one party in the relationship lit a fire under his or her thermostat.

Human Needs

Glasser (1998, 2005) has provided a schema of five individual human motivators signifying the internal nature of human behavior. He sees these as genetic and universal human needs. This array of internal motivators common to all people links humanity together by crossing age, ethnic, racial, and gender differences. The needs are general, not specific, and they include the possibility of many additions and extensions. For example, the third need, power or achievement, could include the ongoing propensity to learn and to attain an increasing amount of knowledge.

Survival, Self-Preservation Most fundamental to human motivation is the need to continue living. The human body seeks to preserve itself in a variety of ways. It digests food, it circulates blood, it shivers when it is cold, it perspires when it is hot, and, without a conscious choice, it inhales and exhales. Human beings develop more conscious and sophisticated survival behaviors in the context of helpful or even threatening circumstances.

As with all the human needs, survival contains interpersonal and social implications. Related to this fundamental need is human sexuality, its urges and yearnings. Consequently, the sex drive plays an important role not only for the pleasure of individuals but also for the continuance of the human race.

Belonging, Love, Affiliation   As human beings grow in maturity, they express their need for connectedness with others. Couples approaching counselors and therapists most often do so because they are unable to satisfy this pervasive need. They feel distant from each other. This estrangement can be expressed by arguing, blaming, criticizing, demeaning, and many other toxic behaviors. No matter what the presenting issue, the reality therapist can usually begin with an exploration of the couple's interactions (i.e., how they treat each other verbally and nonverbally). Through the counseling process, their respective needs for belonging are met and their relationship improves.

Power, Achievement, Inner Control   In general terms, the word *power* has come to imply dominance, exploitation, and even ruthlessness. Yet, as used in the practice of reality therapy, its meaning is rooted in the French *pouvoir* or, in Spanish, *poder*—"to be capable or able." Power is a wider concept than belonging in that it includes many subcategories such as having a sense of inner control, self-esteem, and recognition. People released from hospitals cured of their ailments or injuries feel a sense of power or inner control. And though power can be satisfied by winning a competitive game or sport, it can also be satisfied when a person gains a sense of personal accomplishment irrespective of another person's victory or defeat. The concept of power is broader than that of competition.

In utilizing the concept of power in counseling, the reality therapist takes special note of the couple's effort to satisfy their individual needs by attempting to control the other party. This power struggle often lies at the basis of relationship conflicts. Behavioral symptoms include disagreements about money, sex, lifestyle, and family, as well as an unlimited number of other issues.

Freedom, Independence, Autonomy   Another human motivator is the drive to make choices, to stand on one's own two feet, and to function without undue external constraints. Because of the emphasis on the ability to make choices, Glasser (1998) changed the name of the theory from *control* theory to *choice* theory, thereby correcting a misunderstanding that control meant controlling other people.

No matter how dire the circumstances or how victimized and trapped clients feel, reality therapists help them see that they have at least some ability to make choices and to fulfill their needs more effectively. This book contains examples of clients initially feeling out of control who come to the belief that they have more control than they first perceived.

Fun, Enjoyment   Aristotle defined a human being as a creature that is risible, one that can laugh. A crucial characteristic of mentally healthy people is that they enjoy life. They fulfill their need for fun.

Reality therapists often explore this need by such statements as, "Tell me about the last time you had a hearty laugh." This exploration touches on a deep human quality. Especially effective with depressed clients, this inquiry indirectly teaches clients that there are alternatives to their feelings of sadness, loneliness, and hopelessness. If at one time they were able to enjoy life, even momentarily, it is possible that they can achieve at least some enjoyment in the future.

In summary, the human needs constitute the engines of human behavior. Choices spring from these sources and are attempts to satisfy them. The human needs are general—not specific, universal—not culture limited, innate—not learned, and inevitable in the sense that they motivate all behavior. Reality therapists often use this need schema as an informal diagnostic tool and explore with clients which needs are effectively satisfied or unsatisfied. They then assist clients to make more effective choices for fulfilling their own needs and the needs of other people significant to them.

Quality World and Scales

Emerging from the human need system, individuals develop specific pictures or wants that touch on each need. A want for a relationship with a specific person is based on the need for belonging. A want for success in a particular career connects with the need for power. Engaging in a variety of activities or having a range of choices satisfies the need for freedom, while hobbies and leisure time activities are related to fun. Because these precise wants or pictures are internally satisfying, they are said to have quality.

This collection of precise wants, which includes core values and beliefs, is identified as the *quality world*. It is analogous to a mental picture album. This comparison is also analogous to a drawer containing specific folders. At a given moment, we desire to review the contents of a folder. In other words, we compare what we want, the image of holding a specific folder, with what we have when we are not yet in possession of it. This comparison is also analogous to an out-of-balance scale: We want something but we do not have it. Couples often have many unfulfilled wants. They have scales intensely out of balance regarding their relationship. The reality therapist's task is multiple: help them accept this imbalance, focus on scales that can be realistically put in balance, or develop behaviors aimed at balancing the scales in question.

Total Behavior

Choice theory embraces an uncommon definition of behavior. It refers not only to actions but also to cognition, emotions, and physiology or bodily functioning. Therefore, behavior is a composite. It is total. When the mental scales are out of balance, the mind generates total behavior. Actions are accompanied by thoughts, feelings, and at least some physiological exertion. Reaching into the file drawer involves extending the arm, thinking about how to open the drawer and where to find the file, feelings of hope, and slight changes in physiology. Reality therapists explore the many more complicated behaviors presented by couples: their common activities, their common or conflicted thinking or self-talk, their shared or opposing emotions that focus on their relationship. The skilled practitioner sees these behaviors as choices and treats them *as if* they are choices, even if the couple seems to have little control over specific total behaviors such as feelings of anger, depression, resentment, or indifference.

Perception

Total behaviors chosen by couples are not aimless. Rather, they serve a purpose. All behavior is an attempt to impact our external world for the purpose of gaining something from it. Human beings seek perceptions: the perception of having a satisfying relationship, of a successful career, of being independent or free, of enjoying life. Human beings seek a relationship with another person especially for the purpose of gaining the perception of loving and being loved. Many couples seek counseling because they perceive that their need for love, belonging, connectedness, and affiliation has become strained.

The storehouse of perceptions, analogous to the entire file cabinet, contains not merely wants (i.e., the quality world). It also holds memories, as well as mental projections of future events. These can be pleasant and desirable, unpleasant and painful, or simply neutral—neither pleasant nor painful. A major component of this perceived world is a sense of our degree of personal responsibility for the way we live our lives (i.e., our perceived locus of control).

The skillful reality therapist helps couples explore a significant aspect of their perceptual world—that is, their sense of internal versus external control. In reality therapy, couples identify what they have control over. The reality therapist helps them realize that each of them has control only over his or her own behavior. *We can control only our own behavior, especially our actions and secondarily our thinking* is a foundational principle in choice theory and in the practice of realty therapy.

REALITY THERAPY

The origin of reality therapy preceded Glasser's choice theory (Glasser, 1960, 1965). First developed in a correctional institution for young women and in a mental hospital, it has since been applied to virtually every kind of client, couple, and family who have major or minor problems. It has been used with people from upper and lower socioeconomic levels and a wide variety of cultural and ethnic groups (Wubbolding, 2000, 2011; Wubbolding et al., 2004). Couples desiring remediation from problems such as addiction, abuse, or infidelity, as well as relief from loss, divorce, or posttraumatic stress, seek the help of a competent reality therapist. The development of focus groups wherein couples study applications to their own lives has broadened the application of both choice theory and reality therapy (Glasser & Glasser, 2000, 2007).

Goals of Counseling

The general aim of couples' reality therapy is to help them gain a sense of inner control. When couples enter counseling, they often feel more out of control than they did before making the decision to seek help. They not only feel the pain of their problems but also now feel the pain of having to ask an "outsider" to intervene in their personal business. The initial intervention of the reality therapist should be to congratulate them for taking this step forward. The decision to seek help is thus a step upward—not a step down. More specific goals include the following:

Realize that there are three entities in the counseling process. The reality therapist assists the couple to realize that each person brings unique strengths and limitations to the relationship. And yet there is third entity present: the relationship that exists *between* them. Present in the room are two partners and their relationship itself.

Explore the strength of the relationship. Couples discuss whether their relationship has a slight cold or a terminal illness. They are asked to provide their own assessment and to determine if there is anything in the assessment that they can agree upon. A skilled reality therapist assists them in this process and in describing their common perception of their relationship. Even their agreement about the degree of tentativeness in the relationship can represent a strength.

Gain a sense of need satisfaction. The reality therapist sets a friendly, empathic atmosphere so as to facilitate a therapeutic alliance with the couple. The relationship between counselor and clients facilitates trust, thereby helping clients come to the belief that the reality therapy process will be helpful to them. Building on this trust, the reality therapist assists the couple to gain an enhanced sense of belonging in their relationship. If each person's appreciation for the other increases, they also attain an added sense of power or success. They are encouraged to make choices satisfying to each person—choices that are mutually acceptable and enjoyable.

Change perceptions. In changing actions, couples move from painful to pleasurable perceptions of each other. For instance, if the chemically dependent person seeks help and is supported by the co-dependent partner, they both not only perceive the other person's destructive behaviors but also begin to see their willingness to make changes and their desire to control their own lives more effectively, as well as experiencing an awakening and awareness of the needs of others.

Use quality time. Time spent together is a precise, measurable, and tangible goal. During these special times, the partners choose activities that are need satisfying to both and avoid toxic or deadly behaviors such as arguing, blaming, criticizing, etc. (Glasser & Glasser, 2007; Wubbolding, 2011). This time together strengthens the relationship and creates a storehouse of positive and pleasant perceptions of themselves, their relationship, and each other.

Achieve at least some degree of congruence between quality worlds, behavior systems, and perceptions. The counselor teaches the basic components of choice theory and communicates to the clients that if their relationship is to be need-satisfying to each of them, they will need at least some agreement about what they want from the world around them, including what they want from themselves. They will need to achieve some similarity in how they perceive the world, especially regarding issues of high importance, such as how to raise children or the ability to accept diverse perceptions. Finally, they will reach at least a modicum of commonality in the various components of their behavior: actions, thinking and feelings.

In summary, each goal implies a change in behavior—more specifically, a change in actions. These changes depend upon a willingness on the part of both persons to *want* to improve their relationship and a willingness to *evaluate* their actions and their attitudes followed by a commitment to make alternative *plans*. Change does not occur automatically or by happenstance. It is built on and results from the reality therapist's skill in establishing and maintaining the counseling relationship or therapeutic alliance as well as skillful implementation of the principles of reality therapy that spring from the concepts of choice theory.

The Counseling Process

The artful use of reality therapy occurs when the counselor possesses the skills described in the following discussion, as well as understands the building blocks or philosophical principles of successful counseling: It is a developmental process in that establishing and maintaining trust is a gradual process. When clients feel understood and accepted, there is a high likelihood that they will scrutinize their quality worlds and behavioral systems, as well as their perceptions of themselves, each other, and their relationship. They gradually lessen their defenses and feel comfortable in their self-disclosure and self-exploration.

Clients' Responsibilities   In his lectures, Dr. Glasser facetiously points out that the primary responsibility of clients is to show up. The counselor's responsibility begins at that point. Nevertheless, the success of reality therapy depends on whether clients *want* to change, *want* to improve their relationship and, as trust develops, are willing to make more effective choices. When these conditions are present, clients develop a sense of personal responsibility and a perception of internal control that is expressed by phrases such as "I am willing to …" or "I see your point of view," rather than "You always …," "You never …," and a multitude of other external control statements.

Counselor's Role   The counselor's responsibility is to set an atmosphere that facilitates change. Glasser (1986) states, "The counselor should attempt to create a supportive environment within which clients can begin to make changes in their lives." In developing this idea further, Glasser (2009a, chart) stated that the clients' reconnection "almost always starts with the counselor first connecting with the individual and then using this connection as a model for how the disconnected person can begin to connect with the people he or she needs."

Wubbolding (2008) has identified specific counselor behaviors useful for establishing an appropriate counseling atmosphere:

- Using attending behaviors, including body posture, eye contact, and strategic silence
- Showing accurate empathy (i.e., seeing the clients' points of view)
- Communicating a sense of hope; reality therapy is based on the belief that relationships can always be improved
- Doing the unexpected: continually searching for innovative ways to connect with clients
- Reframing negatives as positives: no matter how bleak the situation, there is always a silver lining
- Discussing problems in the past tense and solutions in the future or present tense because clients subtly learn that they have more choices than previously realized
- Acknowledging clients' feelings without indulging them
- Practicing ethical and professional behavior

Wubbolding describes these behaviors as "tonic behaviors" in that they "enable the client to feel safe, secure and motivated" (p. 373).

Reality therapists set an atmosphere by clearly communicating the division of labor. Clients need to be present and counselors need to disclose their professional credentials, explain the nature of reality therapy, describe the details surrounding duty to warn, informed consent, confidentiality and its limits, and other ethical issues common to all professional relationships (American Counseling Association, 2005; American Psychological Association, 2002; National Association of Social Workers, 1999).

Besides establishing a friendly and engaging atmosphere, the reality therapist intervenes, utilizing procedures that typify reality therapy. In fact, the procedures constitute a methodology for activating choice theory and for helping the counselor connect with clients and assist them to incorporate life-changing choices. These procedures are explained next.

Procedures Utilized by Reality Therapists

Glasser (1972) delineated reality therapy interventions, calling them eight steps. Subsequently, he refined the description of reality therapy to include two major components: establishing the environment and procedures that lead to change. A widespread method for learning reality therapy is the current expression of procedures. Glasser and Glasser

(2008) state, "We now wish to state publicly that teaching the procedures [the WDEP system—wants, doing/direction, evaluation, and planning] continues to be an integral part of training to participants wishing to learn choice theory and reality therapy and is particularly effective in our training programs" (p. 1).

Exploring Quality World Reality therapists help couples to:

- Identify, develop, and clarify their quality world pictures or wants
- Determine their level of commitment (i.e., the therapist helps them explore whether they wish to continue the relationship and how hard they will work at strengthening it)
- Determine their perceived locus of control—how much control they have over their lives and what they can control as well as what is beyond their control
- Explain how they perceive themselves and their respective partners in the relationship

The skilled reality therapist adapts the exploration of wants to individual clients. Some clients are reluctant to respond to direct questioning about their specific wants and needs. Masaki Kakitani (professor at the School of Psychology, Rissho University, Japan, and director of the William Glasser Institute, Japan) translates the question "What do you want?" into "What are you looking for?" or "What are you seeking?" (Wubbolding, 2008).

Prochaska, DiClemente, and Norcross (1992) identified six stages of change related to recovery: precontemplation, contemplation, preparation, action, maintenance, and relapse. Mitchell (2007) broadens the application of these stages to include other resistant clients. In the early stages of change, especially precontemplation and contemplation, clients are not aware of the need for change or experience ambivalence toward change. Mitchell states that clients should be encouraged to discuss their situation in a nonthreatening manner. Consequently, they are more willing to describe how their current relationship is a problem for them rather than being questioned. He adds that especially "leading questions…intended to lead the client toward some insight" are ineffective (p. 74). Thus, the reality therapy procedures are best expressed as explorations rather than merely a series of questions that can feel like an interrogation (Wubbolding, Brickell, & Robey, 2010).

Early in the counseling relationship, the counselor helps clients define their counseling goals or what they hope to gain from the counseling. Do they want a complete cure for the relationship that has a cold or terminal illness? Are they seeking a slight improvement? Do they hope to obtain tools that they can apply in the near and distant future?

Exploring clients' perceived level of commitment means helping them express how hard they wish to work at improving their relationship. Do they wish to have a happy relationship without exerting energy to achieve it? Do they express a moderate level of commitment by such phrases as "we'll try," "we might," or "maybe we can ..."? Do they state that they will do their best or are they at the highest level of commitment expressed by the explicit or implicit statement, "We will do whatever it takes"? The counselor hopes to elicit an agreed upon high level of commitment but often must settle for less, especially in the beginning of the therapy process.

Helping clients discuss their perceived locus of control often entails indirect teaching. Each person learns that he or she can control only one person's behavior: his or her own behavior. Clients discuss how they have tried to control the other person or how they have attempted to coerce or manipulate by arguing, blaming, criticizing, or utilizing the other deadly habits (Glasser, 2000). They also discuss instances when they have successfully focused on changing their own respective actions. The counselor helps them decide which of these choices helps or hinders the relationship.

Part of the initial use of the reality therapy process with couples is helping them present how they see themselves in their relationship. Do they see themselves as part of the problem or do they see only the other person as the problem? Do they express a denial of their own need satisfaction with the motto (stated or unstated), "I go along to get along"? Do they see that their relationship can improve or do they believe it is doomed to failure? How do they see the counselor? Is he or she someone who can help them?

Because of the emphasis on exploring *wants,* the entire process discussed thus far is summarized with one letter: W.

Discussing Total Behavior   Actions, cognition, emotions, and physiological behaviors are the major components of total behavior discussed in descending detail in reality therapy. Because actions are the most controllable component of total behavior, they receive the most direct attention. Accompanying actions are thinking behaviors or ineffective self-talk such as, "I can't do anything right," "The situation is hopeless," or "Nobody can tell me what to do." Effective self-talk includes "I am competent. I can," "My life will improve," and "I am happiest when I am generous in my relationships." Included in total behavior are emotions or feelings. Although emotions such as anger or depression are

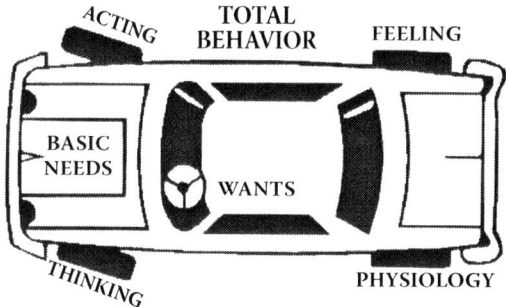

TOTAL
ACTING BEHAVIOR FEELING

BASIC
NEEDS
WANTS

THINKING PHYSIOLOGY

Figure 1.1 Total behavior.

sometimes the most obvious component of total behavior, we have less direct control over them than over actions and cognition. Nevertheless, counselors do not ignore them. They are analogous to the lights on the dashboard of an automobile. When they light up, they signal a message to the driver: "Something is wrong; you need to take action."

Even physiology occupies a place in the use of reality therapy. Stress and strain, anger, and resentment, as well as tolerance, serenity, and joy, have various effects on physiology. However, when clients change actions and cognition, they also alter feelings and, to a lesser degree, their physiology. Glasser illustrates total behavior as the four wheels of an automobile: The front wheels are action and thinking; the rear wheels are feelings and physiology (Figure 1.1). When the driver of the car accelerates and turns the steering wheel, the front wheels direct the car and the back wheels follow. In human behavior, there is a time lag between changing actions and changing feelings. The feelings do not simultaneously follow a change in actions.

Finally, the implementation includes dealing with every component of human behavior. For pedagogical reasons, the four aspects of total behavior can be summarized in one word and one letter: doing—*D*.

Conducting Self-Evaluation The successful use of reality therapy is based on counselor interventions focusing on helping clients evaluate their own behaviors. It is like the keystone in an arch holding the structure together. If it is removed, the arch crumbles. When clients feel they can trust the counselor, they are more likely to conduct a searching and fearless self-inventory—an evaluation of their behavior. This assessment is founded on the concept of whether their respective behaviors are improving or damaging the relationship. Counselor interventions include such statements as the following:

- Describe how you helped or hindered your relationship in the last day, week, and month.
- Elaborate on behaviors you bring to this relationship that enhance it.
- How do you communicate with each other in ways that bring you close to each other? In ways that damage the relationship?

Self-evaluation also includes a self-judgment on the attainability of quality world wants (Wubbolding, 2000). Specific interventions are:

- Discuss whether your expectations of each other are realistically attainable.
- Will you be able to gain what you want from this relationship?
- Share with us your thoughts about how need satisfying this relationship will be in the future.

The reality therapist assists the couple to evaluate their own behaviors (not the behaviors of the other party), the attainability of their wants, and whether their own actions are bringing them closer together or farther apart. He or she also helps couples evaluate their degree of commitment to their relationship as well as where they see their control (i.e., their perceived locus of control). Behavioral change and improved relationships occur and develop only after the parties involved judge that current choices are helping or not helping. These self-evaluations, along with estimates of want and need fulfillment, occupy at a central place in the use of reality therapy. The arch of reality therapy remains strong and effective when these interventions spring from a therapeutic alliance characterized by deep mutual trust, confidence, empathy, concern, and positive regard.

Additional self-evaluation interventions focus on level of commitment, such as:

- Tell me how your current level of commitment will help you improve or prevent you from improving your relationship.
- Describe what you mean by "trying" and whether trying is sufficient.

Still other self-evaluations center on perceived locus of control:

- Describe the impact on yourself and your partner of trying to regulate his or her behavior.
- We have discussed toxic or deadly behaviors as well as tonic or caring behaviors. When you have used them, what impact have they had on your relationship?

Glasser (2009b) strongly reemphasized the central place of self-evaluation in the practice of reality therapy. In his clinical demonstration at the Evolution of Psychotherapy, an international conference sponsored by the Milton Erickson Foundation, Glasser counseled a woman who was considering a divorce. In many ways, her marriage was beneficial and desirable. She was affluent, secure, and married to a famous personality. Yet she firmly believed he was having an extramarital affair, so she was ambivalent about staying in the marriage. Throughout the 50-minute session, Glasser repeated his theme idea:

> The key issue for you is to determine whether you are better off in the relationship or out of it. ... I suggest you ask yourself, "Am I better off with him or without him?" ... Try to determine whether the benefits of the marriage to him outweigh the negatives or vice versa. ... Which is better for you: to stay or to leave?

Using the cost/benefit technique, he assisted the client to evaluate her quality world, her desire to save the relationship. She needed to judge which course of action was more attractive. Only with this self-evaluation could she resolve her feelings of ambivalence, mistrust, and uncertainty. By his own actions, Glasser taught that self-evaluation lies at the heart of effective reality therapy.

In summary, self-evaluation constitutes the sine qua non for the successful implementation of the principles of choice theory and its delivery system reality therapy. The reality therapist assists couples to examine the helpfulness of their wants or quality world, the various components of their total behavior, and their viewpoints or perceptions. The many types of self-evaluation can be summarized by the letter *E*.

Making Plans   Plans agreed upon by couples are essential to the practice of reality therapy applied to relationships. The saying "to fail to plan is to plan to fail" summarizes the planning component of the reality therapy process. Efficacious plans spring from the wants and goals of the couple. Planning occurs only after clients have, at least in a seminal fashion, expressed their desire to improve the relationship and have decided that their current choices are not improving the relationship or are unsatisfying to them. The goals of couple counseling provide guidelines for effective planning:

- Strengthening the relationship. Ideally, the couple agrees to take action that would bring them closer together.
- Gaining a sense of need satisfaction. Couples formulate plans that are individually satisfying. Plans formulated mutually are

therefore more likely to be habitual. If either individual recoils from the plan or is even indifferent toward following through on it, the likelihood of such choices becoming consistent is lessened.

- Changing perceptions and quality time. Reality therapists know that plans mutually agreed upon are most helpful to them. Choosing an activity performed without arguing, blaming, complaining, and criticizing draws the individuals together and allows them to build a storehouse of perceptions favorable toward each other. These pleasant memories provide a foundation for future resolution of disagreements, compromises, and negotiation.

The desired result of alternative actions explicitly chosen by the couple equates with congruence in their wants (quality worlds); in their feelings, thoughts, and actions (total behavior); and in their viewpoints of their respective world views (perceptions).

The full achievement of these goals resulting from action plans is ideal and rarely perfect. Couples often come to realize that the components of their choice system (i.e., wants, behaviors, and perceptions) will never be a perfect match. Consequently, a question that reality therapists need to help clients come to grips with is not only "What do you want in your relationship?" but also, more realistically, "What will you settle for?"

SUMMARY

Choice theory provides a comprehensive system for understanding human motivation. It is based on the viewpoint that human beings control their own lives. They are not victims of their childhood, their family interactions, or their cultural environment. They choose most of their behaviors and are capable of altering them. As Glasser (2008) has said, "We need not be victims of our past or our present *unless we choose to be*" (p. i). Reality therapy is the delivery system making choice theory operational, practical, and usable by professionals and by clients. Implementing the WDEP system of reality therapy can be life changing and has been shown to be validated (Wubbolding, 2011) and applicable to a multiplicity of clients from various cultures and ethnic groups with a wide variety of presenting issues.

REFERENCES

American Counseling Association. (2005). *Code of ethics*. Alexandria, VA: Author.
American Psychological Association. (2002). *Ethical principles of psychologists and code of conduct*. Washington, DC: Author.

Glasser, W. (1960). *Mental health or mental illness?* New York, NY: Harper & Row.

Glasser, W. (1965). *Reality therapy.* New York, NY: Harper & Row.

Glasser, W. (1972). *The identity society.* New York, NY: Harper & Row.

Glasser, W. (1986). *The basic concepts of reality therapy.* Chatsworth, CA: The William Glasser Institute.

Glasser, W. (1998). *Choice theory.* New York, NY: HarperCollins.

Glasser, W. (2000). *Reality therapy in action.* New York, NY: Harper Collins.

Glasser, W. (2005). *Defining mental health as a public health issue.* Chatsworth, CA: The William Glasser Institute.

Glasser W. (2008, Summer). Reality therapy teaches. *The William Glasser Institute Newsletter,* p. i.

Glasser, W. (2009a). *How the brain works* (chart). Chatsworth, CA: The William Glasser Institute.

Glasser, W. (2009b, December). *Clinical demonstration of reality therapy.* Presented at the Evolution of Psychotherapy International Conference sponsored by the Milton Erickson Foundation. Anaheim, CA.

Glasser W., & Glasser, C. (2000). *Getting together and staying together.* New York, NY: Harper Collins.

Glasser W., & Glasser, C. (2007). *Eight lessons for a happier marriage.* New York, NY: Harper Collins.

Glasser W., & Glasser, C. (2008, Summer). Procedures: The cornerstone of institute training. *The William Glasser Institute Newsletter,* p. 1.

Mitchell, C. (2007). Effective techniques for dealing with highly resistant clients. Johnson City, TN: Mitchell Publishers.

National Association of Social Workers. (1999). *Code of ethics.* Washington, DC: Author.

Powers, W. (1973). *Behavior: The control of perception.* New York, NY: Aldine.

Prochaska, J., Di Clemente, C., & Norcross, J. (1992). In search of how people change. *American Psychologist, 47,* 1102–1114.

Wiener, N. (1948). *Cybernetics.* New York, NY: John Wiley & Sons.

Wiener, N. (1950). *Human use of human beings.* Boston, MA: Houghton Mifflin.

Wubbolding, R. (2000). *Reality therapy for the 21st century.* Philadelphia, PA: Brunner Routledge.

Wubbolding, R. (2008). Reality therapy. In J. Frew & M. Spiegler (Eds.), *Contemporary psychotherapies for a diverse world* (pp. 360–396). Boston, MA: Houghton Mifflin Company.

Wubbolding, R. (2011). *Reality therapy: Theories of psychotherapy series.* Washington, DC: American Psychological Association.

Wubbolding, R., Brickell, J., Imhof, L., Kim, R., Lojk, L., and Al-Rashidi, B. (2004). Reality therapy: A global perspective. *International Journal for the Advancement of Counseling, 26*(3), 219–228.

Wubbolding, R., Brickell, J., & Robey, P. (2010). A partial and tentative look at the future with a practical idea. *International Journal of Choice Theory and Reality Therapy, 29*(2), 25–34.

2

AN INTERVIEW WITH WILLIAM
AND CARLEEN GLASSER

Patricia A. Robey and Robert E. Wubbolding

INTRODUCTION

Dr. William Glasser is often reticent to talk about his past and his personal life, noting that "where I am now is more important than how I got there, but others may be interested in the journey" (Roy, 2006, p. 55). Glasser was born in 1925, the youngest of three children. He became aware at an early age that there was tension in his family. In the book *Choice Theory* (1998), Glasser expressed his admiration and love for his father, Ben, but noted that he had some difficulty with his mother, Rebecca. Rebecca had an urge to control others and rule the marriage, which set the tone in their home. Even as a child, Glasser recognized that his parents' marriage was not healthy for either of them, and he was a witness to events in the household that frightened him. Fortunately, when Glasser was 6 years old, Ben Glasser gave up his resistance to Rebecca's control and the fighting between his parents stopped. It is likely that these early experiences influenced William Glasser's view of human nature and the behaviors he adopted to cope with the important people in his life.

Glasser married Naomi Silver in 1946 and had what he described as a good marriage until Naomi died of cancer in 1992. After Naomi died, Glasser was very lonely and looked for a new companion to fill the gap that Naomi's death made in his life. At the same time, Carleen was going through a very difficult divorce. Because Carleen was an instructor for the William Glasser Institute and had been friends with Naomi Glasser, Bill Glasser knew her well. However, it was not until a mutual friend suggested that Glasser should get in touch with Carleen that they came together as a couple. The match was a success and they have been married since 1995. Together they have authored four books on relationships: *The Language of Choice Theory* (1999), *Eight Lessons for a Happier Marriage* (2007), *What Is This Thing Called Love? The Essential Book for the Single Woman* (2000b), and *Getting Together and Staying Together: Solving the Mystery of Marriage* (2000a). Carleen wrote,

> He's taught me to let the painful past go and not blame anyone, including myself. ... Commiserating over past failure is a waste of precious time that could be used to create a more satisfying present. Bill has taught me how to be happy. (Glasser & Glasser, 2000a, p. 23)

The preface of *Choice Theory* begins,

> This book is about how important good relationships are to a successful life ... if we are not sick, poverty stricken, or suffering the ravages of old age, the major human problems we struggle with—violence, crime, child abuse, spousal abuse, alcohol and drug addiction, the proliferation of premature and unloving sex and emotional distress—are caused by unsatisfying relationships. (Glasser, 1998, p. ix)

Choice Theory provides an explanation for why human beings struggle in relationships and also provides concrete direction for the types of behaviors we can choose to have happier relationships. Glasser advises us to stop trying to control and change others, especially the important people in our lives. We do this by eliminating what Glasser calls the seven deadly relationship habits: criticizing, blaming, complaining, nagging, threatening, punishing, and bribing or rewarding to control. Instead, we practice the seven caring relationship habits: listening, supporting, encouraging, respecting, trusting, accepting, and always negotiating disagreements (Glasser & Glasser, 2000a).

In this interview with Patricia Robey and Robert Wubbolding, the Glassers talk about their early experiences in relationships and how these experiences influenced the development of Glasser's work, and they share how the ideas presented in *Choice Theory* have influenced the happiness they share in their relationship together.

INTERVIEW

Patricia Robey: I would like to begin by thanking you very much for agreeing to this interview with me and Bob. As you know, we are working on a book that addresses how the ideas of choice theory and the practice of reality therapy can be used in counseling couples. We thought it would be interesting to hear from you, not necessarily just about choice theory and reality therapy, but also about your own experience with working with couples and about your own personal stories.

Carleen Glasser: Well, we both have interesting sets of parents.

Robey: We could start there!

William Glasser: My parents' relationship was terrible. My father and mother went into extremely violent fights while I was still young because my mother would taunt him and taunt him and taunt him until he couldn't stand it anymore. But I've forgotten most of that because my father was the best father a man could ever have and he put up with my mother and I know a lot about him that I want to tell. It doesn't make any difference. That's the way it was.

Robert Wubbolding: Bill, what was it about their relationship that influenced you?

William Glasser: It's hard to say. They had a poor relationship, my father and mother.

Carleen Glasser: Would you say that their relationship was based upon extreme external control? Basically, your mom trying to control your dad for the whole of their lives?

William Glasser: My mom tried to control everybody, and that included my father.

Carleen Glasser: But she was good to you, wasn't she?

William Glasser: Well, my mother was a good mother to me.

Carleen Glasser: It was just, you know, inevitable that trouble was going to come. There are just people who are like that and they are very destructive to relationships. In those days, people didn't get divorced. Bill tells me that his dad finally just gave up and went along with whatever she [Bill's mother] said. In their later years, they got along okay because he stopped fighting her external control.

In my own parents' relationship, it was very similar. My father was the controlling one. My father was the boss. My mother was Italian, and she was very emotional and very gentle and very artistic. But she was the one who had to comply with whatever his wishes were. I can remember my dad saying about me and my sister and my mom that you three women are against me and you're all plotting secrets with one another. A lot of secrets went on in our house because we were afraid to say anything to him because he would blow up and scream and holler.

He wasn't violent; he didn't beat us or hit us or hurt us in any way. But he liked to holler, and when you have a 6 foot something huge man and he's the only male in the house and he hollers a lot, and he's very loud and very strong willed, and three little women, two little girls and a tiny woman who's my mother, we felt intimidated. So we had to create our own little culture of, you know, you can't tell dad this. How are we going to tell dad that? We would always go to my mom and ask her what we should do and she didn't know. Consequently, the good news is I didn't turn out like either one of them, because I saw a model that I didn't want to follow in my own life. I think Bill created choice theory with this huge emphasis on how external control destroys relationships and we both experienced that first hand.

So when Bill and I got married we decided we will have no external control in our marriage with each other—none. And we would be supportive and loving and all the things our parents weren't to us. We learned by process of elimination, I guess you could call it, and I think our marriage has been extremely successful because, well, number one—we were more mature, and we both knew choice theory. I knew choice theory and he taught me what I didn't know about choice theory by example, by being a choice theory person, and his way of dealing with me was far better than I had ever been dealt with by any other man in my entire life.

Robey: Talking about early recollections leads me to think about first dates and those types of things, as you start to move in connecting relationships with others. How did you start dating and how were you thinking about relationships at that time?

Carleen Glasser: Well, I wanted to get out of the house as soon as I could because my dad was so overpowering, controlling, and everything. So I married at a very young age a man who was 7 years older than me and had my first child at the age of just barely 20. That was my way of doing things as a solution. I went to a year of college and after that year I got married. It got me out of the house; from the frying pan to the fire, unfortunately.

When my son went to kindergarten, I went back to college. I finished my degree and I became a teacher, and I started a whole professional life. It changed my whole persona. I became so much more.

Robey: You got out of the house, but then things changed for you when you started to develop your own identity.

Carleen Glasser: Well, I grew up. I started dating my first husband when I was 15, and what did I know? I knew nothing. I was a kid and he was older and he was really, really handsome and it just, you know, we grew up! I grew up and grew apart and got more education. I went my own way and ended up with my professional career and everything—getting involved in counseling and getting involved in reality therapy, and ended up after many years finally finding Bill. My gosh, that was something that I never expected to happen, but it was probably the best thing that ever could have happened to me—getting involved and learning the reality therapy from Bob [Wubbolding], learning about Glasser, being on the [William Glasser Institute] board of directors eventually, and seeing Bill, and then getting to know Naomi. His wife Naomi and I really liked each other. I knew Naomi better than I knew Bill! All the experience I had throughout the years led me to be ready to be there for Bill when he needed me. I see a relationship as people finding each other when they need each other. He needed me and I needed him. He had lost Naomi. She died suddenly of cancer and we all grieved and mourned her loss in the institute, me included, and about a little over a year later, he was very lonely, I was available—it just all worked out, like it was like meant to be. And so, I was there for him, he was there for me. We needed each other and I think when there is a need present in life you find some way to fulfill it. He was already in my quality world because of all this studying of his ideas and everything. I think in some small way, I might already have been in his quality world, too.

Wubbolding: Bill, tell us some of your early recollections.

William Glasser: When I went to college, I went to Case Institute of Technology. I didn't care for that at all and I didn't pay any attention and I didn't study at all, I just hung around the fraternity and I got C+ grades. There weren't any very good grades, but I made up my mind that I wanted to be a psychologist, and I even started taking psychology before the war ended. I was drafted into the army, even though there was no fighting, and I spent 8 months in the army. I got the GI Bill to finish up everything in college, and in college I started learning psychology. I figured it was obvious that psychology was nothing then; you had to become a medical doctor to become a psychiatrist. So I applied for medical school.

Carleen Glasser: Were you married to Naomi then?

William Glasser: Yes, I married her when I was 20 or 21 years old. Naomi went to the army with me. We stayed near Salt Lake City. By that time, I knew what I wanted to do. I wanted to go to medical school. So, I really learned how to study when I finally got into medical school. I loved medical school. It was a wonderful experience because I always hated college and there I loved it. I was married and I was getting along with my wife all right.

Carleen Glasser: So can you tell us how you met Naomi and what it was like being married to her?

William Glasser: I met a girl named Audrey and I really fell in love with this girl but … anyway, she told Naomi that I was a good guy. Audrey really cared for me; she was in love with a sailor and when he came back from the navy, I was really upset because I knew she was going to go back with him and that wasn't any fault of mine. He came back from his service and I was dating her while … I wasn't sleeping with her or anything like that; I was just dating her. In those days you didn't go to bed immediately with people you knew like that! At least I didn't!

Wubbolding: Times were different!

William Glasser: Anyway … so when he came back, then she suggested that I go out with Naomi and so I got involved with Naomi then. Then we got married and ultimately I went to medical school. Well, by this time, I had been studying and working on choice theory and I wrote my first book. I was really excited when that was published because I didn't know how to get a book published. I was in psychiatric practice; I was seeing a few people and I used to go home for lunch. Everything was close by. Anyway, I went to the book store, the used-book store. I really counted all the books and in that day, the most books were by Harper and Collins. I started writing books and Naomi helped me with that to some extent, even though for various reasons we didn't get along so well together then.

Carleen Glasser: You had a good working relationship with her in terms of the books and everything.

Wubbolding: Bill, tell us how you and Carleen came to be a couple. What do you think the secrets or the guidelines are for a happy marriage?

William Glasser: When I married Carleen, I was really kind of looking for someone that I really could devote my life to.

Carleen Glasser: After Naomi died, you started dating because you were lonely. I remember you writing a letter to me and saying that you were not involved with anyone and would I be interested in coming to see you.

William Glasser: But we didn't get married right away.

Carleen Glasser: Well, we didn't get permission from our children!

[laughter]

Carleen Glasser: I was working with my school district, but I was teaching in another school in the city of Cincinnati, the Schwab Middle School. I was teaching how to do choice theory in their classrooms and I told them that I was dating the guy who wrote the book and that was Bill Glasser. They didn't believe it! So he came and talked to them. The Cincinnati Public School System asked if he would work with this school and he said, well, I will if you hire Carleen as a quality school instructor/quality school consultant, which they did. I spent my last year in education, my 25th year, working with Bill Glasser in a school. We worked that whole year together and it was great.

William Glasser: And the school was successful. Then I wrote something in the book that got us married.

Carleen Glasser: Oh, that was the funniest part! I was working at the Schwab School in the office and he was at home writing the book *Staying Together*. So he is finishing up *Staying Together* and he has everything finished, but HarperCollins said, "We need an end flap." A blurb for the end flap … they say we need it and we needed it yesterday. So he called me up at work and said HarperCollins wanted the end flap and I've been working on writing it all morning, can I read it to you?

William Glasser: Yeah, I read it to her.

Carleen Glasser: He read it to me and the end of whatever he said about the book, he said, "Carleen and I will be married in the summer." That was the first I heard of it! He was very romantic. I started to cry.

William Glasser: Then I sent you the letter too. I sent a letter …

Carleen Glasser: That was just to get us started … the letter you sent and then I took about 2 weeks to answer the letter at the onset of our relationship and he ... I found out later that he rarely picks up his mail. He will wait for 2 to 3 days to see ...

William Glasser: If you don't pick it up, no bad news!

Carleen Glasser: I found out later that he went every day to the mail-box to look for my letter of response. You still have it in your drawer. I found it.

Wubbolding: You didn't give him any flak did you?

Carleen Glasser: I said no. I gave him a lot of flak! I said no, I have to wait for a certain amount of time and think about this, and finally I said OK. Then we courted ... we talked on the phone several times. I wanted to kind of get to know what his inten-tions were—whether they were honorable or dishonorable, and they were rather quite dishonorable!

[laughter]

Carleen Glasser: And I thought, well, being an instructor in the insti-tute, I didn't want to be a topic of gossip and ridicule and all of that and be dumped shortly after and then what would hap-pen when I retire from the school district that I was working with? I planned to start teaching reality therapy workshops and intensive weeks, and I thought what would happen to my career? I had a lot of things to mull over. So I was thinking for 2 weeks and finally he got my answer and I said OK in October. It was like in August when he wrote to me. I said in October, "Let's ... I have some time and maybe we can get together." That's why on Halloween he flew in and we got grounded by this snow storm in Cincinnati, then the next day flew to New York and we had a wonderful time. We found out we are so compatible.

Wubbolding: So it sounds from what you said that common interest, doing things together—that's what kept you together? And that's what cemented the relationship?

Carleen Glasser: We worked together. We both were working together with choice theory and reality therapy.

William Glasser: I worked 1 or 2 days a week and then I traveled mostly one time a week.

Carleen Glasser: And then he would write. He would write and I would be reading what he had written and editing. I mean, we spent the first 10 years of our marriage writing 10 books. He wrote and I edited and some of them we wrote together. So we were busy. I think being busy, having the same things to do ...

Robey: We are actually drawing close to our time to finish up. Before we do, I would like to ask you, Bill, what you would like to tell people about relationships that they need to know.

William Glasser: Well, two things in relationships, and Carleen knows them. First of all, leave each other alone!

Carleen Glasser: Let each other be; it's like the Beatles song. Not alone, alone like never see each other, but leave each other be; in other words be who you are.

William Glasser: Because if I'm working and Carleen wants me to eat lunch … she goes, "Come to lunch, come to lunch, come to lunch …" It didn't bother me. I just kept working. I didn't come to lunch; that's all.

Carleen Glasser: I learned by process of elimination to let him be. But also, I take saying, "Let each other be" to mean even beyond that; be who you are, acceptance.

William Glasser: By that time, I had pretty much known about the deadly habits and things like that.

Carleen Glasser: We never used the deadly habits.

William Glasser: I never use the deadly habits with anyone. I never criticize, never blame.

Robey: So you said the one piece of advice was let it be, let each other be. And the second one?

Carleen Glasser: Never use the deadly habits.

William Glasser: Never criticize … I reduced it to CBC: Never criticize, blame, or complain. Those are the most important things.

Carleen Glasser: My secret to a happy relationship is like we said in the book, *What Is This Thing Called Love?* The definition of love is commitment—being willing to make a commitment to the person that you are with.

Wubbolding: I gather from what you said about "not only let it be" but there has to be some overlap in your quality world: what you want and what you do and how you spend your time.

Carleen Glasser: Well, everything that he wanted to do was in my quality world. And not everything that I want to do is in his quality world because I like to go shopping and he doesn't mind shopping but I like to also do painting and drawings and sculptures and things like that. He appreciates what I do.

Wubbolding: Well, thanks a lot, Bill, and thanks, Carleen. We appreciate your time and look forward to sharing your story with our readers!

REFERENCES

Glasser, W. (1998). *Choice theory: A new psychology of personal freedom.* New York, NY: HarperCollins.

Glasser, W., & Glasser, C. (1999). *The language of choice theory.* New York, NY: HarperCollins.

Glasser, W., & Glasser, C. (2000a). *Getting together and staying together: Solving the mystery of marriage.* New York, NY: HarperCollins.

Glasser, W., & Glasser, C. (2000b). *What is this thing called love? The essential book for the single woman.* New York, NY: HarperCollins.

Glasser, W., & Glasser, C. (2007). *Eight lessons for a happier marriage.* New York, NY: HarperCollins.

Roy, J. (2006). *Development of the ideas of William Glasser: A biographical study.* Doctoral dissertation. Retrieved from ProQuest Dissertation and Theses. (Accession Order No. AAT 3227052).

II

Issues and Applications

3

MULTICULTURAL COUPLES
Seeing the World Through Different Lenses

Kim Olver

INTRODUCTION

With the development of technological communication, access to international travel, and increases in immigration, it seems that societies are becoming more and more diverse. These changes in the way that we interact with one another allow us to experience people from different cultures, religions, races, ethnicities, and nationalities (Samovar & Porter, 1995). As a result of these developments, as well as changes in the taboos against interracial and intercultural couples, it is not surprising that the number of intercultural couples is increasing. In fact, "a record 14.6% of all new marriages in the United States in 2008 were between spouses of a different race or ethnicity from each other. ... That figure is an estimated six times the intermarriage rate among newlyweds in 1960 and more than double the rate in 1980" (Passel, Wang, & Taylor, 2010, p. ii).

According to Passel et al. (2010), the taboo against interracial marriage was changed in 1967, when the U.S. Supreme Court ruled in *Loving v. Virginia* that a Virginia statute barring whites from marrying nonwhites was unconstitutional. Since that landmark decision, the number of interracial marriages has dramatically increased, from 65,000 in 1970 to 422,000 in 2005. Even so, it took years for the public to accept interracial dating, with only 48% of those polled in 1987 agreeing that this was acceptable. As societal attitudes changed, however, those numbers grew to 83% by 2009.

In spite of this change, research indicates that differences in culture create stress on relationships that could be addressed in therapy (Bhugra & DeSilva, 2000; Heller & Wood, 2007; Hsu, 2001; McFadden & Moore, 2001; Molina, Estrada, & Burnett, 2004). For counselors, one of the most challenging issues can be counseling the multicultural couple. Challenges can occur not only between individuals in the couple, but also between the counselor and the person whose culture is dissimilar to his or hers (Roysircar, Arredondo, Fuertes, Ponterotto, & Toporek, 2003). Conflict, judgments, and disapproval can also come from parents, children, extended family members, friends, and society at large (Bustamante, Nelson, Henriksen, & Monakes, 2011).These challenges illustrate the essential role of cultural experiences in determining our worldview.

CULTURAL EXPERIENCES

According to Olver and Baugh (2006),

> We are often unaware of our culture because it is so much a part of who we are; we simply think of our culture as "normal" … Often our total unquestioning acceptance of our culture may tell us that people who do things differently are "wrong" and not simply "different." (p. 1)

Whenever two people embark on a committed relationship, they already bring with them two separate and distinct cultures. When variables such as different ethnicities, religions, generations, socioeconomic status, or other cultural variables are added, those cultural differences are amplified. Hall (1976) stated that "culture is man's medium; there is not one aspect of human life that is not touched and altered by culture" (p. 16).

Acculturation

Acculturation occurs when contact between two cultural groups results in cultural changes to both groups (Berry, 1998). Differences in levels of acculturation and desire for acculturation can impact the couple's ability to manage relationship stressors. For example, a first-generation immigrant who is not acculturated to his or her new country may have more difficulty in adapting to an intercultural relationship with a partner who is more acculturated to the social norms in which they live.

Our acculturation process begins at birth. Once we are taken home from the hospital or experience a home birth, we begin acculturating to our immediate surroundings. This begins within the home where we live, extending to a small circle of influence and then to the broader community. According to choice theory (Glasser, 1998), throughout this process we are accumulating knowledge from the information and life experiences we have. Later, as we perceive information from the *real world,* we compare our perceptions of the information we are receiving with what we already know, in an effort to make sense of the world. Over time, we develop values and beliefs and decide whether information and experiences are pleasurable, painful, or simply neutral. Together, what we know and what we value—what Glasser (1998) refers to as our *total knowledge filter* and *valuing filter*—determine how we see and experience the world. Given the unique nature of these two filters, it becomes obvious that an individual's values and perceptions can be vastly different from other people's values and perceptions. Furthermore, the greater the degree of dissimilarity of their individual cultural experiences is, the greater the discrepancy between their values and perceptions can be.

Values and Beliefs

Multicultural couples are likely to face challenges due to the diversity of their values, beliefs, and attitudes (Hsu, 2001). Understanding and accepting this diversity requires a willingness to communicate and understand the other's cultural perspective (Waldman & Rubalcava, 2005). Forming one's values and beliefs throughout the acculturation process is a natural developmental process. Unfortunately, what often accompanies one's values and beliefs is a sense of ethnocentrism, or the belief that the values and beliefs that have been adopted are unequivocally correct, instead of one's unique interpretation of the world. A sense of righteousness ensues, which then can lead to conflict between individuals who do not share the same interpretations.

For example, consider the case of someone who comes from a family where women stay home until their children are school age to provide the best possible care for their children. This person may be judgmental about other women who do not stay home because she believes her position is correct and not simply an opinion or perception of what is best. Conversely, someone who comes from a background where a woman's professional contributions are valued may believe that any woman who stays home with her children is wrong for not accomplishing her full potential. Neither position is unequivocally correct, but rather simply involves a cultural value that is "right" for the person who holds it. It is not an unequivocal truth for all people.

Individualistic Versus Collectivistic Cultures

Different worldviews are notable between individualistic and collectivistic cultures. According to Brooks (2008),

> The individualistic countries tend to put rights and privacy first. People in these societies tend to overvalue their own skills and overestimate their own importance to any group effort. People in collective societies tend to value harmony and duty. They tend to underestimate their own skills and are more self-effacing when describing their contributions to group efforts. (paragraph 6)

These differences in worldview can have a major impact on the couple relationship, as well as the counseling process. From the perspective of choice theory (Glasser, 1998), a person from an individualistic culture is often socialized to value power and freedom, while a person from a collectivistic culture might value survival and love and belonging most highly.

CHALLENGES OF THE MULTICULTURAL COUPLE

Like many couples, multicultural couples experience challenges related to how to communicate, roles assigned by gender, parenting styles, etc. These challenges, however, may be exacerbated due to the profound differences in the worldviews and behaviors that are indicative of cultural values, experiences, and beliefs (Bustamante et al., 2011; Wehrly, Kenney, & Kenney, 1999). Bustamante et al. (2011) found that cultural values related to time and family connections also created stress for intercultural couples. Couples stated that their relationships with their spouse's family members were strained and that they often felt excluded and marginalized when in the presence of in-laws.

When one or both partners in the relationship take the position that their way is the "right" way, increased tension in the relationship often follows. When differences present themselves in the couple relationship, individuals may respond with curiosity about how the other person views the world while seeking understanding, or they may become self-righteous about their own perceptions and view their partner as misguided or wrong. When couples respond with curiosity and a desire for understanding, their relationship can grow. Conversely, when couples respond with righteousness, their relationship can become strained.

Implications for Counseling

It is a counselor's responsibility to be very clear in his or her own values and perceptions. If he or she begins to see a situation from a "right" or "wrong" perspective, then the counselor also knows to switch to curiosity, seeking understanding from the person whose perceptions are different. Effective counseling requires counselors to learn to empathize with the ways their clients see their worlds, acknowledging and respecting these differences. Further, to minimize the influence of their own cultures in the counseling process, counselors must develop their own multicultural competence (Henriksen, Watts, & Bustamante, 2007; Roysircar, Arredondo, Fuertes, Ponterotto, & Toporek, 2003; Sue, Arredondo, & McDavis, 1992).

Brandell, McRoy, and Sebring noted:

> Counselors are often unaware of the unique cultural effects of race and ethnic background on the development of identity and it seems that it is essential to possess skills in assessment and intervention in family and organizational dynamics related to racial identity issues if mental health counselors want to achieve optimum effectiveness. (as cited in Solsberry, 1994, paragraph 3)

Similarly, when the counselor finds himself or herself believing that one member of the couple is correct based on the counselor's worldview, triangulation can develop.

The overemphasis on or denial of race as an influence in a relationship can be an indication that a counselor has biases related to these issues. Extensive focus on issues of race may cause the counselor to lose focus on the clients themselves (Solsberry, 1994). Therefore, it is crucial for the counselor to maintain the balance of exploring cultural issues while not falling prey to the use of culture to explain everything that is occurring in the relationship.

CASE STUDY

Eric, an African American, 40-year-old male, and Kathy, a 36-year-old Caucasian female, sought help with their relationship. They had been in a committed relationship for 6 years, living together for the past 4 years. Both stated that they had no interest in marrying and were content with the status of their relationship.

Kathy had made the appointment and began talking about what she perceived as the problem in their relationship. She reported being frustrated by what she perceived as Eric's insensitivity to her feelings.

He said that she did not understand the pressure he was under to be a black man "dealing with" a white female. She then accused him of being ashamed of her, asked if he possibly were involved with another woman, and wondered why he wanted to keep their relationship secret. Eric responded by saying that there was no other woman and that he was simply a private person and what he did in his personal life was no one else's business.

As the story unfolded, it became clear that Kathy and Eric were having difficulty understanding each other because of their different cultural frameworks. Neither could actually "hear" the other person because each could not make sense of the experiences and perceptions of the other.

The counselor asked them if they would like to try the first caring relationship habit (Glasser & Glasser, 2000) of *listening*. She explained they would take turns telling each other their stories; the other person's job was to listen and attempt to understand the speaker's point of view. They were not expected to reach consensus. Their assignment was simply to do their best to understand each other's point of view. This involves adopting the attitude of curiosity rather than judgment. They were to ask questions until they understood the other person's position, thus developing a deeper understanding of the total knowledge and valuing filters of each other. Once a greater understanding was achieved, they would be better equipped to work toward negotiation. They agreed to follow through on this suggestion.

Whenever a counselor uses reality therapy and choice theory with couples, it is beneficial to prevent either partner from criticizing the other. Glasser (2000) stated, "Any marital counseling that allows one partner to blame the other will harm the marriage" (pp. 37–38). Therefore, the counselor may, at times, have to take a strong role to ensure that individuals honor the process of listening without interruption.

Eric's Story

Eric began by describing himself as a strong, proud black man who did motivational speaking for audiences he described as "pro-black," black people who are vocal about their strong racial pride. Since his involvement with Kathy, however, he believed that he was being perceived by many of his friends and acquaintances as a fraud. After all, how could he be pro-black and involved with a white woman? Eric even reported losing work contracts because of his connection to Kathy. People who had hired him previously refused to contract with him when they heard that he was involved with a white woman. However, he recognized the importance of their relationship. He said that he loved Kathy and did not want their relationship to end.

Eric reported growing up with his parents in a predominantly black community. When he was 8 years old, his family moved to a predominantly white neighborhood. He said that as a preteen living in a white community, he had no color consciousness until introduced to music performed by black entertainers. This is when Eric began identifying with his specific black culture and its differences from white culture. From that point on, Eric began seeking to understand and clarify his identity as a dark-skinned, African American male who was larger in stature than most of his peers. This clarification continued through a move with his family to Mississippi and continues today.

Eric said that to be a successful African American male in the United States was an active process and he did not want to lose status in the African American community because of his involvement with a white female. When questioned, he still declared himself as pro-black in every aspect, but his struggle occurred when most people would not even ask the question of why he was with a white woman. They simply made judgments, such as that he no longer liked black women, that white women were better than black women, that his message of empowerment for blacks had been watered down, or that he was somehow not quite as strong as he was previously and could be easily challenged.

As a result, Eric chose to go into predominantly black environments without Kathy because he did not want to deal with the stress of what others said about their relationship. He believed that he had to be particularly on guard when he was in predominantly black circles with Kathy. He explained that it was possible that black men would challenge him and perceive him as weak. He was also concerned that black females might disrespect Kathy or him because many women he knew and respected were angry with him for crossing the racial line. Because he was from a collectivist culture, the perceptions and feelings of others were extremely important to Eric.

When Eric had finished speaking, Kathy was able to repeat back to him the main points he had made and she worked to understand what he was experiencing as an African American man. While she admitted that she did not completely understand him, she believed that she had made great progress in understanding him better and was no longer feeling the need to argue her point of view.

Kathy's Story

When it was Kathy's turn, she explained how hard it was for her to be excluded from events that were a large part of Eric's life. She loved him and wanted to share in all things that were important to him. She felt unimportant and discarded when he went places without her—especially places to which he would normally want to take his woman if she were black. Kathy believed that, by hiding the fact that they were a couple, Eric was indicating that he was ashamed of her.

She explained that friends and family had asked her if Eric was married or involved with another woman because of his unwillingness for their relationship to be public. Because he was so private about his personal life, she was not sure whether or not he was involved with someone else. She believed there were occasions when he at least spent time with other women but did not tell her about it. That frustrated her. She wanted him to trust her and not withhold important information.

When Eric socialized without her, Kathy felt saddened, disrespected, and abandoned. She wanted him to understand and appreciate her feelings and added that she exerted much effort to see his point of view while he made little effort to see hers. She also believed that he minimized her complaints and saw them as mere attempts to manipulate him into doing what she wanted him to do. Other people's perceptions of their relationship were of little concern to her. What mattered most to her was their relationship, which she believed was being undermined by Eric's concern for everyone's feelings except hers.

When Kathy was finished, Eric was able to repeat the main points she had made, demonstrating his understanding of her position.

Once they each had their turn, the counselor asked them to talk about what was good about their relationship. Both agreed that they had a lot in common. They did similar work, both had an interest in music, and they shared similar values. Eric said theirs was a relationship without stress from within the couple. He said he had no pressure or drama from Kathy. The only pressure he experienced was from people outside their relationship—mainly his friends and people with whom he worked. He appreciated Kathy's giving nature. Kathy said Eric was exactly the man she had always dreamed of. He was strong but not overbearing, confident but not arrogant, and a great communicator.

Individual Sessions

Eric said that he was happy with the relationship as it was. It was Kathy who seemed to be frustrated. So the counselor spent the remaining sessions with Kathy, individually, teaching her choice theory. Glasser (1998) stated that "in marriage, as in all human relationship problems, someone has to take initiative and stop using external control" (p. 177). In this instance, Eric was not attempting to control Kathy, but Kathy was trying to control or change Eric.

During her first individual session with Kathy, the counselor began by teaching her one of choice theory's basic premises: the only behavior that Kathy could control was her own. Intellectually, Kathy realized this, but she began to recognize all the ways in which she was attempting to get Eric to recognize her publicly. She pouted, did a lot of deep sighing, and would withdraw into herself in a way that Eric noticed. Kathy used all of these behaviors unsuccessfully to induce Eric to notice her distress and change his behavior.

According to Olver (2011), there are always three options in a relationship: change it, accept it, or leave it. Not wanting to leave the relationship, Kathy had been trying unsuccessfully to change it for a long time. She loved Eric and wanted them both to be happy. Therefore, she decided she wanted to work on accepting it. Accepting meant giving up all anger, frustration, and resentment associated with the things she wished were different (Olver, 2011).

Appreciation

The next step was to work on appreciation. According to Olver (2011), accepting is a good place to get to in relationships but appreciation is even better. Olver proposed that those things that bother one partner about the other are actually good for the other person when he or she is willing to take the time to look for the benefits. Therefore, Kathy was given the homework assignment to create a list of how Eric's troublesome behavior actually had some hidden positive value for her.

When Kathy was able to let go of her self-imposed rule that couples *should* include each other in all things, she realized that Eric's behavior of doing certain things independently actually benefited her in several ways. Kathy realized that if Eric were to include her in all things, she would have no time to herself. Her business would suffer because of the amount of time spent at his events. Also, she realized that if Eric felt stress when he included her, their relationship could be seriously strained. She did not want to cause unnecessary stress for him. She also recognized that Eric's behavior was providing her with an opportunity to practice unconditional love.

Once she created this list and could honestly practice appreciation instead of tolerance or acceptance, Kathy and the counselor decided it was time to terminate sessions, with occasional follow-up as needed.

Kathy called one month after termination to say that things were going very well. Because a couple is a system, it is notable that as Kathy became more accepting of the situation, Eric also began to make changes. While still concerned about the perceptions that others have of him, he has independently decided to include her in most of what he does professionally and socially. Using choice theory and reality therapy, he found a way to prioritize what was most important for him. Glasser (2003) stated, "In a contentious marriage … even if only one partner starts to use Choice Theory, the other will be hard pressed to continue using external control" (p. 81).

DISCUSSION

All relationships experience some degree of multicultural issues when two individuals decide to become a couple. No two people have identical cultural experiences; there are always differences in their respective total knowledge and valuing filters. However, with interracial and multicultural couples, there are additional stressors—some due to cultural differences and others due to their perception of society's response to their relationship. Eric and Kathy experienced multiple cultural differences between them that they were both working to understand better. Additionally, there was interference from people outside the couple who impacted Eric's ability to make his living. This was a reality that could not be ignored. Other couples also deal with judgments from parents, children, friends, and extended family members.

In the case of Eric and Kathy, the most profound difference was the fact that Kathy was raised in an individualistic culture and Eric in a collectivist culture. Eric had great concern for his family and friends, who did not understand his choice of a partner. He wanted to protect their feelings without disappointing them. Kathy, who was from an individualistic culture, learned the value of doing what made her happy and expected everyone else to adjust to her choices. If people could not adjust, they were not true friends. Understanding and appreciating these differences helped Kathy and Eric stop trying to change each other, which ultimately saved their relationship.

Nonjudgmental Neutrality

A skilled counselor must be careful not to "take sides" and to work to understand the cultural values that differ from his or her own. Wubbolding (2000) wrote, "Reality therapy … needs to be adapted to the perceptions and cultures of the clients. … The therapist needs to respect the family's culture and adjust the use of reality therapy to it" (p. 83). In addition, "Reality therapy, artfully applied, provides a bridge connecting therapists to individuals in their uniqueness and diversity" (Wubbolding, 2011, p. 112).

Counselors have their own cultural lenses, which color their perceptions of the world. Astute counselors who work with multicultural clients in any capacity also work to uncover their own biases and prejudices. This is not as easy as it sounds because their total knowledge and valuing filters are completely "right" for them. Consequently, it is sometimes difficult for them to conceive that there could be other ways, equally correct, for people raised in other cultural environments. Considering a perspective different from one's own as valid can be extremely challenging.

Supervision

Whenever counselors find themselves thinking that one person in the relationship is right and the other is wrong, it might be a good time to seek supervision and, whenever possible, to consult a colleague who has a better understanding of the culture of the client who the counselor thinks is "wrong." Counselors' cultural values cause them to see the world from their unique perspectives. A supervisor or consultant skilled in multicultural issues can help counselors process their own biases and assist them to return to a neutral, objective position. Glasser (2000) stated, "Any marriage counselor who even intimates that one partner is more responsible than the other for the problem risks doing further harm to the marriage" (p. 40).

Structured Reality Therapy

One of the strengths of using choice theory with couples is the model of structured reality therapy, which can be used with couples in which both partners are interested in improving their relationship (Glasser, 1998). It involves a list of six questions asked of each individual that allows both people to hear from their partner what is wrong with their relationship as well as what is good about it. After receiving that information, each person makes a commitment to do one thing every day for the next week that will benefit the relationship. There is no pressure on the individual to choose anything. Neither partner feels coerced into making the change. They both recognize that it is something they want to do for the good of the relationship. Theirs is a gift freely given that is internally motivated out of a strong desire to improve the relationship.

When a couple commits to doing something significant to improve their relationship—and they follow through every day for a week—something magical happens. The dynamics of the relationship improve almost immediately.

If the couple is unable to convince the counselor that both partners want help for their relationship, then structured reality therapy will not work (Glasser, 1998). In that case, it is best to work with the person who is most unhappy about the relationship to help him or her individually find a way to change his or her behavior, accept the behavior of the partner, or leave the relationship in order to alleviate his or her unhappiness.

SUMMARY

If a client comes from a culture where it is acceptable to tell one's partner what he or she wants in the couple relationship, then help the client develop the skills and confidence needed to do so. People have the right to ask for what they want in a relationship. In this instance, when people are unhappy, they should tell their partner by describing the problem and requesting a change. However, partners do not always honor these requests—not because they do not love their partner, but rather simply because the behaviors they are already using work for them.

When one person in the relationship is not getting what he or she wants and the partner has been asked for change and has not granted it, then the unhappy partner must decide whether or not this is non-negotiable. The same holds true when the unhappy partner exists in a relationship where it is not acceptable to ask directly for what he or she wants in the partnership. There are basically three options in a relationship: change it, accept it, or leave it (Olver, 2011). If the behavior is non-negotiable, the choice may be to leave. Some examples of non-negotiables are infidelity, gambling, domestic violence, child abuse, and chronic financial irresponsibility (Olver, 2011). Non-negotiables are different for each individual. However, if there has not been a violation of a non-negotiable and the unhappy person wants to stay in the relationship, then it is the counselor's task to help the client figure out how to change his or her own behavior or to accept the partner's behavior.

Teaching Kathy how to appreciate the formerly problematic behavior in Eric helped her to change her perception, which then changed the whole experience for both of them. Consequently, now there is a balance between what she wants and what she gets from the relationship.

Choice theory helps people understand that "different" does not mean "wrong"; it just means different. When individuals can truly engage the seven caring relationship habits (Glasser & Glasser, 2000) of listening, supporting, encouraging, trusting, respecting, accepting, and negotiating differences in their relationship, they can begin to appreciate, instead of judge, the other person, even if they are unable to come to agreement. With mutual understanding come a greater willingness to negotiate and a determination to find a win/win/win solution (Olver, 2011). Both individuals get what they need and the relationship grows stronger as a result of going through the process of prioritizing the relationship over each person's individual needs.

REFERENCES

Berry, J. W. (1998). Acculturation and health: Theory and research. In S. S. Kazarian & D. R. Evans (Eds.), *Cultural clinical psychology: Theory, research, and practice* (pp. 39–57). New York: Oxford University Press.

Bhugra, D., & DeSilva, P. (2000). Couple theory across cultures. *Sexual and Relationship Therapy, 15*, 183–192.

Brooks, D. (2008). Collective vs. individualistic societies. *The New York Times.* Retrieved from http://legacy.signonsandiego.com/uniontrib/20080814/news_lz1e14brooks.html

Bustamante, R. M., Nelson, J. A., Henriksen, R. C., Jr., & Monakes, S. (2011). Intercultural couples: Coping with culture-related stressors. *Family Journal, 19*(2) 154–164. DOI: 10.1177/10664807113997233

Glasser, W. (1998). *Choice theory: A new psychology of personal freedom.* New York, NY: HarperCollins.

Glasser, W. (2000). *Reality therapy in action.* New York, NY: HarperCollins.

Glasser, W. (2003). *Warning: Psychiatry can be hazardous to your mental health.* New York: NY: HarperCollins.

Glasser, W., & Glasser, C. (2000). *Getting together and staying together: Solving the mystery of marriage.* New York, NY: HarperCollins.

Hall, E. T. (1976). *Beyond culture.* Garden City, NY: Anchor Press/Doubleday.

Heller, P. E., & Wood, B. (2007). The influence of religious and ethnic differences on marital intimacy: Intermarriage versus intramarriage. *Journal of Marital and Family Therapy, 26*, 241–252.

Henriksen, R. C. Jr., Watts, R. E., & Bustamante, R. (2007). The multiple heritage couple questionnaire. *The Family Journal, 15*, 405–408.

Hsu, J. (2001). Marital therapy for intercultural couples. In W. S. Tseng, & J. Streltzer (Eds.), *Culture and psychotherapy: A guide to clinical practice* (pp. 225–242). Washington, DC: American Psychiatric Press.

McFadden, J., & Moore, J. L. (2001). Intercultural marriage and intimacy: Beyond the cultural divide. *International Journal for the Advancement of Counseling, 23*, 261–268.

Molina, B., Estrada, D., & Burnett, J. A. (2004). Cultural communities: Challenges and opportunities in the creation of "happily ever after" stories of intercultural couplehood. *Family Journal: Counseling and Therapy for Couples and Families, 12*, 139–147.

Olver, K. (2011). *Secrets of happy couples: Loving yourself, your partner, and your life.* Chicago, IL: InsideOut Press.

Olver, K., & Baugh, S. (2006). *Leveraging diversity at work: How to hire, retain and inspire a diverse workforce for peak performance and profit.* Chicago, IL: InsideOut Press.

Passel, J. S., Wang, W., & Taylor, P. (2010). Marrying out: One-in-seven new U.S. marriages is interracial or interethnic. *Pew Research Center Publications.* Retrieved from http://pewsocialtrends.org/files/2010/10/755-marrying-out.pdf

Roysircar, G., Arredondo, P., Fuertes, J. N., Ponterotto, J. G., & Toporek, R. L. (2003). *Multicultural counseling competencies 2003: Association for Multicultural Counseling and Development.* Alexandria, VA: Association for Multicultural Counseling and Development

Samovar, L. A., & Porter, R. E. (1995). *Communication between cultures* (2nd ed.). Belmont, CA: Wadsworth.

Solsberry, P. W. (1994). Interracial couples in the United States of America: Implications for mental health counseling. *Journal of Mental Health Counseling 16*(3), 304–317.

Sue, D. W., Arredondo, P., & McDavis, R. J. (1992). Multicultural counseling competencies and standards: A call to the profession. *Journal of Counseling & Development, 70,* 477-486.

Waldman, K. & Rubalcava, L. (2005). Psychotherapy with intercultural couples: A contemporary psychodynamic approach? *American Journal of Psychotherapy, 59,* 227–245.

Wehrly, B., Kenney, K. R., & Kenney, M. E. (1999). *Counseling multiracial families.* Thousand Oaks, CA: Sage.

Wubbolding, R. (2011). *Reality therapy: Theories of psychotherapy series.* Washington, DC: American Psychological Association.

Wubbolding, R. E. (2000). *Reality therapy for the 21st century.* Philadelphia, PA: Brunner Routledge.

4

RELATIONSHIP RECOVERY AFTER INFIDELITY

Patricia A. Robey and Maureen Craig McIntosh

INTRODUCTION

Perhaps no problem is as common in counseling couples as infidelity. From presidents to sports figures, from film stars to the neighbor down the street, most of us know about the trauma that results from an affair. In Western cultures, when people enter into a committed relationship, there is usually an understanding that the relationship will be monogamous (Kaslow & Hammerschmidt, 1992) and that affairs are shameful and unforgivable (Brown, 2007). This understanding and the meaning that couples attach to monogamy are socially constructed (Atwater, 1979). The values and beliefs about faithfulness in relationships are influenced by religious edicts and supported by the legal system. As a result, individuals are socialized to believe in the concept of faithfulness and the restriction of sexual behavior to a committed relationship.

On the other hand, contradictory messages are delivered through media sources, which glorify sexuality without regard to marital status (Atwood & Seifer, 1997). From a choice theory perspective, the picture of monogamy is one that most people hold as important and is part of the quality world picture of the ideal relationship. It is not surprising to almost anyone that an act of infidelity can be traumatizing to a relationship.

Partners respond to affairs with guilt, dishonesty, lies, anger, humiliation, depression, anxiety, and regret. In some cases, infidelity can have an impact on careers, reputations, family, and loss of time and money (Lusterman, 2005; Vaughan, 2003). For some, there may have been a suspicion of infidelity with a denial from the offending partner. When the infidelity is confirmed, the betrayed partner loses trust in the offender and begins to question every communication. Even past treasured memories are likely to be reexamined and colored by doubt (Lusterman, 2005).

Betrayed partners may also begin to question their own role in creating relationship deficits leading to affairs. They blame themselves for their perceived inadequacy in the relationship, thus leading the partner to stray (Pittman & Wagers, 2005; Vaughan, 2003). They feel that they have lost their identity as being part of a couple, feel a loss of connection with their partner, and even a loss of purpose, as they strive to discover a new sense of personal identity in relation to the unfaithful partner (Spring, 1996). Physiologically, affairs can result in sexually transmitted disease, unwanted pregnancy, abortion, sexual dysfunction, and even suicide or homicide (Humphrey, 1987; Vaughn, 2003).

WHY PEOPLE CHEAT: INFIDELITY AND BASIC NEEDS

Glasser (1998) explained that all human beings are born with basic needs—genetic instructions that include love and belonging, power, freedom, fun, and survival. All behavior serves the purpose of satisfying one or more of these basic needs. People have very specific ideas of how to get these needs met in relationships, and they develop pictures of what their ideal relationships will look like. These pictures, which are socially constructed, become part of what Glasser refers to as *quality world* pictures.

Relationships would be much easier and would have less conflict if we all shared the same pictures of and attitudes about sex and intimacy (Rasmussen & Kilborne, 2007). The challenge for a couple is that the partners often have different pictures of how to get needs met in their relationship. For example, one partner may prefer a one-to-one intimate connection to satisfy the need for love and belonging, while the other might meet his or her need for belonging through relationships with groups of people. Without effective communication regarding what they feel is lacking in the relationship, partners only know that something is wrong and that they are unhappy.

Love and Belonging

Glasser (1998) noted that the drama related to love and belonging is the stuff from which great literature is made. According to Glasser, we have a genetic need for love and belonging, but we are not programmed for how to satisfy that need. Compounded with the need itself is that it requires attention for a lifetime. There is no genetic instruction that requires us to be with the same person forever: "Our genes want someone; they don't care whom" (Glasser, 1998, p. 34). Unfortunately, the desire and attraction initially felt toward our partner often fades over time. When love begins to fail, it most often fails because we begin to use controlling habits in an effort to get our partner to change.

Interestingly, some unfaithful partners argue that their infidelity serves the purpose of helping the relationship (Atwood & Siefer, 1997). One argument is that affairs are a normal part of marriage, are good for the marriage, and can actually revive a marriage that has lost its spark. Another rationale is that the desire to improve the marriage results in infidelity with the intention of creating jealousy, thus increasing intention from the faithful partner. More often, unfaithful partners blame others for the deficits in the relationship. Infidelity is seen as a response to the couple's unhappiness and means that love is gone from the relationship and separation is imminent. The familiar complaint that "my partner doesn't understand me" may be an excuse for behavior, but may also express an unfulfilled need for acceptance, love, friendship, and intimacy in the relationship.

Power

Power is related to sexuality because of the desire to see ourselves replicated in our offspring. It is also related to the desire to conquer, to win, to control, and to have things go the way we want them to go. People with high needs for power may sacrifice anything—or anyone—in their efforts to get it. Power drives us to succeed and attain the physical trappings in life that show we have met our goals. Unfortunately, people with high needs for power often use coercion to control others. As Glasser (1998) wrote, coercive power destroys love: "No one wants to be dominated, no matter how much those who dominate protest their love" (p. 42).

Power is also associated with respect. In the couple relationship, respect means being cared about, heard, attended to, and desired. When partners do not have that, they may initially struggle to get it. Eventually, though, people get tired of the struggle, give up, quit communicating, and disconnect from their relationships. Glasser and Glasser (2000) suggested that it is not usually a lack of love that destroys relationships. Rather, it is the power struggles between partners that provide the biggest obstacles for long-term happiness.

Some of the issues related to power and infidelity include retaliation for perceived or real injuries to the couple relationship, a desire to hurt or challenge one's partner, rebellion, or a desire for conquest (Atwood & Siefer, 1997). People find power in affairs because they feel compensated for feeling inadequate, inferior, or vulnerable in other areas in their lives. For example, when people feel unattractive or concerned that they are aging and unable to perform, they may blame their partners for the problem and seek reassurance in an affair. Finally, in some circles, having numerous affairs may be a sign of social conquest and social status. In some instances, affairs may even be a side effect of the process associated with negotiating and conducting business.

Freedom

The concept that humans are genetically programmed for freedom implies trouble for relationships. Beginning with engagement, the process of committing to a relationship with a ring denotes that one is promising to be bound to the other for life. People desire to have choices and to feel free to come and go when they want. When we agree to join with another person, in essence we compromise our own freedom for the sake of the relationship. Glasser and Glasser (2000) noted that the only way to deal with freedom needs in a couple relationship is through negotiation. In fact, the Glassers suggested that couples often separate because they feel that their freedom is compromised—only to enter another relationship and experience the same restrictions.

Issues related to freedom and infidelity include the desire to experience new approaches and variety in sexuality, such as experimentation with nudity, sexual techniques, and sexual positions. People who want freedom in their sexual lives often feel sexual fatigue due to the confinement of career, household, and family responsibilities. The frustration they feel in being trapped in their relationships results in a need to assert independence, escape, and emancipation from conventional standards. When both partners feel the same way, some construct rationales that support alternative lifestyles, such as approval for extramarital affairs, mate-swapping, or "swinging."

Fun

Of all the needs, fun should be the easiest to satisfy in relationships (Glasser & Glasser, 2000). However, the Glassers noted that many people find it difficult to have fun in their couple relationships. The Glassers hypothesized that the reason that people stop having fun together is related to the use of deadly relationship habits (criticizing, blaming, complaining, nagging, threatening, punishing, bribing, or rewarding to control) in other areas of the couple's lives together. When couples are struggling with power or freedom issues, often the last thing that comes to their minds is to create opportunities for fun together.

While it can be argued that all of the aforementioned rationales for infidelity can provide much needed fun for the unfaithful partner, there are a few that seem to suggest the fun in having sex with a new partner. First, the new lover is often considered to be more sexually attractive or better than the old partner. Infidelity can relieve the boredom that comes with lack of learning and experimentation in the couple's sex life. Finally, sex can provide a quick fix for the desire for pleasure and recreation that may be missed or neglected in the primary relationship (Atwood & Siefer, 1997; Pittman & Wagers, 2005).

Survival

Survival of the species is based on sex, but the drive for sexual activity is based on pleasure as well as survival. Pheromones, which are secreted by the body, are responsible for sexual attraction. These chemicals are very powerful and are designed to bring us together to reproduce. Of course, love and sexual desire are not the same thing. The sexual arousal is so powerful and pleasurable that people often confuse it with love:

> Driven by survival hormones that care nothing about love, many people are willing to act as if they are in love when they are not … they have sex for pleasure with people they don't even like, much less love. The sex feels good for one or both, and that becomes sufficient reason to have it. (Glasser, 1998, p. 35)

Problems related to sexual survival in relationships include infrequent, absent, or poor intercourse in the relationship. Some argue that having sex with a partner outside the relationship might actually "cure" sexual dysfunction. Others suggest that infidelity can overcome issues such as depression, neurosis, and other choices that some would refer to as mental illness (Atwood & Siefer, 1997), but that people who understand choice theory regard as behavioral choices that deal with unhappiness. Other sexual behaviors related to survival are clarifying sexual orientation, using sex to become pregnant and thus carrying on the species, and earning money through selling sexual services in order to support one's family or lifestyle (Atwood & Siefer, 1997).

CASE STUDY

Tracey and Dan were a Caucasian couple in their mid-30s. Tracey was an executive for an international banking corporation and Dan was the owner of a small construction business. They had been together for 6 years and had married when Tracey became pregnant with their daughter, Natalie, who was 2 years old now. It was the first marriage for both.

Dan made the initial call requesting counseling. In the intake, Dan stated that he and Tracey were experiencing stress in their relationship but did not specify the nature of the stress. Only Dan arrived for the first session. Dan reported that Tracey was reluctant to attend counseling and had only agreed to come at Dan's insistence. At the last minute, Tracey claimed that she had to go out of town on business and told Dan he could cancel the appointment or go on his own. Frustrated, Dan decided that it would be in his best interest to attend the session and get his story heard.

Dan needed little encouragement to tell his story. He said that he and Tracey had been happy together until the birth of Natalie, which seemed to stifle them both. "Don't get me wrong, we love Natalie," Dan said. "It's just that we can't seem to find time to spend alone any more, to travel, or even go to the movies." When asked about their sex life, Dan replied, "Well that's part of the problem. We almost never have sex anymore."

The story unfolded. Three years earlier, Dan read an e-mail message from one of Tracey's colleagues. The e-mail made suggestive remarks about Tracey and referred to the fact that the writer and Tracey had shared a room and engaged in sexual behavior together while Tracey was on a business trip. "I was furious," Dan said. When he confronted Tracey, she said that the message was just a joke—wishful thinking on the part of her colleague. However, Dan's suspicions were aroused. He began to follow Tracey, listened in on her phone calls, checked her e-mails, and questioned her behavior. Eventually, Dan's suspicions were confirmed when he saw Tracey and her lover exiting a motel room and kissing one another good-bye.

Tracey became pregnant. At Dan's insistence, she had a paternity test, which indicated that Dan was the baby's father. The couple decided to make a commitment to each other and to the baby and got married. Unfortunately, they each found it difficult to put the past behind them. "I can't get the thought of Tracey having sex with that guy out of my mind," Dan said. The counselor asked if Dan wanted to try to save the relationship. When he said yes, the counselor said that she would like Dan to encourage Tracey to come to the next session as well. The counselor and Dan worked together to plan a strategy for how Dan could invite Tracey in a tactful way, rather than to coerce her into coming to counseling.

Tracey did come to the next session, although she appeared reluctant and somewhat skeptical. Tracey's story was similar to Dan's except that Tracey expressed her anger with Dan's coercion and blamed him for initiating the problem between them.

The counselor spent some time empathizing with both Dan and Tracey and complimented them on the courage they displayed in coming to counseling. The counselor then asked the couple whether they were coming to counseling because they wanted to make the relationship better or whether they were there to end it. The couple seemed surprised at this question. "I thought your job was to fix us," Tracey said. The counselor then said that this was not possible. She explained that it was up to the couple to decide if they were willing to do the required work, though the counselor would do her best to help them create new strategies for their relationship that would bring them closer together. The counselor then helped the couple assess the strengths that they would bring to counseling and define their expectations and desired goals for counseling and their relationship.

As the sessions continued, the counselor helped the couple to understand that the infidelity, while serious in its own right, was also a symptom of other issues in their marriage. She made them aware of how the ways that they communicated and treated one another were a major influence on the disconnection they were feeling in their relationship. The counselor taught them about the concepts of caring and deadly relationship habits (Glasser & Glasser, 2000) and gave them a list of these habits that they placed on their refrigerator at home as a reminder of how they wanted to interact with one another. The counselor worked with the couple to examine the pictures that they had of how they wanted their relationship to be. They found that they both had idealized and unrealistic pictures of the marital relationship. They had been trying to fit themselves into the stereotypical pictures of what a couple should look like. To be a good wife and mother, Tracey felt that she should have stayed home with Natalie. To be a good husband and father, Dan felt that he should be the person who would support the family financially.

Tracey realized that she had a high need for freedom and resented being "trapped" into the marriage by her pregnancy. She did not enjoy being a mother and having to tend to the needs of her daughter. She wanted to travel, to pursue power on her job, and to be answerable to no one but herself. Dan, on the other hand, had a strong picture of the couple acting as a unit. For him, that meant that they would do everything together with Natalie. The counselor helped them to create new pictures of how they could be a unit and yet each could meet his or her individual needs within the relationship. They renegotiated their roles as partners and as parents. Dan took more of the responsibility for Natalie's care, which he enjoyed and which gave him the sense of family belonging that he craved. Tracey was able to focus on her job, which gave her the freedom she wanted. Together, they agreed that they would focus on having satisfying times together when Tracey was available.

It is not uncommon for partners to have differing beliefs about sexuality and intimacy (Rasmussen & Kilborne, 2007). The counselor encouraged Dan and Tracey to discuss their sexual relationship and the values and beliefs that they held about sexuality. Withholding sex is sometimes perceived as lack of sexual desire. However, it may also be a symptom of a power struggle, with partners holding on to past resentments. Dan and Tracey stated that they initially enjoyed sex with one another, but that intercourse had indeed become part of a power struggle that was used as punishment and reward. The counselor asked, "If your relationship were the way you wanted it to be, would sex be an issue?" The couple said it would not, and they recounted the early stages of their relationship when they enjoyed a great deal of sexual activity. The counselor asked about the couple's use of medications, street drugs, and alcohol, noting that all can have an effect on desire and performance. When Dan and Tracey assured her that their sexual relationship was not influenced by substance use, the counselor gave them resources and discussed activities that would help bring them closer together while they were working on the other challenges they faced in their relationship.

The counseling seemed to be going well until the seventh session, when both entered the office in a rage. Tracey had gone out of town on business again and Dan had reverted to his former behavior of nagging, blaming, and criticizing. Tracey responded by defending her position and pointing at Dan as the source of their problems. After listening to each complain about the other for a brief time, the counselor asked them to stop and breathe quietly for a moment. When they had themselves under control, the counselor began to process what had happened. Understanding that there was a purpose behind Dan's behavior, the counselor asked, "Dan, what was going on with you that got you so upset when you heard Tracey was going out of town?" Dan replied, "It just brought up all those old memories for me. I can't get over it. I just feel like I can't trust her." With tears of frustration, Tracey replied, "How can I earn your trust if you don't let me out of your sight? It seems like everything we've worked for these last few weeks has been a waste of time and money."

The counselor helped Dan to realize that his lack of trust was based on his fear that he would be hurt again in the relationship. "Well, how can I get over this?" Dan asked. Tracey said, "I'm not willing to give up my life so that you don't have to worry." It seemed that they were at an impasse.

The counselor asked them again to decide whether they wanted to remain married or if they wanted to end the relationship. They both renewed their commitment to working on it. Tracey said, "My parents were divorced and I hated it. I don't want that to happen for Natalie." Dan agreed. The counselor then asked each to consider how his or her behavior was working for or against the relationship. They both agreed that what they were doing was not helping them to become closer. "It's hard, though," Dan complained. The counselor agreed: "Unfortunately, I can't make a miracle for you," she said. "But what I can tell you is that if you keep using coercive behavior with one another, there is a good likelihood that your relationship will fail. It's up to you."

DISCUSSION

After an affair, partners often have difficulty in restoring the trust and intimacy they previously shared. They struggle to decide whether they will try to save the relationship and move forward or whether they will separate (Atkins, Baucom, & Jacobson, 2001; Vaughan, 2003; Whisman, Dixon, & Johnson, 1997). If partners decide to stay together, they must begin to reestablish a means of communication and behavior that will help them move forward toward recovery. As we saw in the case of Tracey and Dan, the partner who feels betrayed may begin to view the betrayer through a microscope, questioning the partner's activities and wondering what the partner is doing and whether the partner is secretly meeting the third party. Rather than bringing the couple closer together, however, Dan's effort to ease his worry had the countereffect of distancing him from Tracey. He was so worried about getting hurt that he chose relationship habits that kept him disconnected from Tracey. Dan put up a wall in the relationship that served the purpose of protection, but ultimately did nothing really to protect the relationship. Tracey became defensive and felt pressured to provide detailed reports on all her activities. As a result, a power imbalance created another sense of disconnection in the relationship.

As the couple began to work on re-creating the relationship, it was natural that they might feel some fear. In counseling, they had to address difficult issues and problems. Because there was already a breach of trust, it was unlikely that they would want to risk the vulnerability required to communicate thoughts and feelings. The barrier in communication further intensified the issues present in the relationship and increased the difficulty in coming to a resolution regarding them.

Dan and Tracey came to three more counseling sessions before terminating the counseling relationship. While they initially had made great efforts to rebuild their relationship, it began to appear that the couple was unlikely to put aside their own interests for the best interest of the couple relationship. As Wubbolding (1988) wrote, "A crisis occurs when one or both members ... want the other person to match his or her own pictures and *is unwilling to change this want*" (p. 92). A few months after the counseling ended, however, the counselor received a card from the couple in which they wrote that they were still together and were continuing to work on their relationship.

According to Glasser and Glasser (2000), partners need to be friends in order for a relationship to succeed. Friendship relies on having equal power, which is based on listening to one another and negotiating conflicts. In their efforts to defend their positions, Dan and Tracey became embroiled in a power struggle that threatened to override their individual needs and the needs of the relationship. As Glasser (1998) wrote, "When there are differences, as there have to be the longer you know each other, you must work them out to stay in love. When you can't, you are no longer in love" (p. 164).

IMPLICATIONS FOR COUNSELING

Infidelity as the presenting problem in counseling is not uncommon, with an estimated 50%–65% of couples in therapy dealing with issues related to infidelity (Atkins, Baucom, Eldridge, & Christensen, 2005; Atkins et al., 2001). Therapists find adultery to be one of the most difficult problems to treat in therapy. Only physical abuse has been found to be more damaging to relationships (Whisman et al., 1997), and research suggests that, at the conclusion of treatment, couples still felt distress as a result of an affair (Kessel, Moon, & Atkins, 2007). Pittman and Wagers (2005) noted that there are myths related to infidelity that inhibit couples from discussing the problem. These myths include the notion that infidelity is normal and can be expected, that divorce is a natural consequence of an affair, that the faithful partner is lacking and therefore at fault for the betrayal, and that keeping the infidelity a secret will somehow save the relationship.

When counseling couples with issues of infidelity, counselors need to be aware that the view of infidelity is a social construct that usually is viewed as problematic, if not deadly, for the relationship. Counselors, too, may share this view and tend to approach the issue from a problem-based point of view. However, focusing on the affair may actually amplify the problem and lead the couple to believe that the marriage cannot survive (Atwood & Seifer, 1997).

From a choice theory perspective, the behavior of infidelity is considered to be purposeful and need satisfying. Therefore, the counselor views behavior as clients' best attempts to get their needs met, although the behavior chosen may be less effective than an alternative behavior. Defining behavior in this way helps to neutralize the counselor's personal response to the situation, thus creating a nonjudgmental environment in which the counselor and clients can work on creating a successful outcome. Some questions that might be considered include: What needs were met by the infidelity that were not being met in the relationship? What values and beliefs were compromised by the affair? What roles have the couple assumed that influence how their relationship is governed? What can be learned from this experience that will help the relationship to move forward?

Working with issues of infidelity requires counselors to have an open mind and to be flexible in the clinical choices they make (Hertlein, Wetchler, & Piercy, 2005). Counselors must be aware of their own quality world pictures related to couple relationships and fidelity:

> Inevitably, therapists will have to address questions concerning their own values and beliefs about affairs: Are affairs always wrong? If there has been an affair does that mean there must be a separation or divorce? Can someone be in love with more than one person at one time? (Nelson, Piercy, & Sprenkle, 2005, p. 190)

Counselors should be aware that what is problematic for one couple may not be a problem for another (Linquist & Negy, 2005). The relationship can actually be stronger after an affair, if clients choose to do the necessary work (Brown, 2007; Nelson et al., 2005).

REFERENCES

Atkins, D. C., Baucom, D. H., Eldridge, K., & Christensen, A. (2005). Infidelity and behavioral couple therapy: Optimism in the face of betrayal. *Journal of Consulting and Clinical Psychology, 73*(1), 144–150.

Atkins, D. C., Baucom, D. H., & Jacobson, N. S. (2001). Understanding infidelity: Correlates in a national random sample. *Journal of Family Psychology, 15*(4), 735–749.

Atwater, L. (1979). Getting involved: Women's transition to first extramarital sex. *Alternative Lifestyles, 2*, 33–38.

Atwood, J. D., & Seifer, M. (1997). Extramarital affairs and constructed meanings: A social constructionist therapeutic approach. *American Journal of Family Therapy 25*(1), 55–75.

Brown, E. M. (2007). The affair as a catalyst for change. In Peluso, P. R. (Ed.), *Infidelity: A practitioner's guide to working with couples in crisis* (pp. 149–165). New York, NY: Routledge.

Glasser, W. (1998). *Choice theory.* New York, NY: HarperCollins.

Glasser, W., & Glasser, C. (2000). *Getting together and staying together: Solving the mystery of marriage.* New York, NY: HarperCollins.

Hertlein, K. M., Wetchler, J. L., & Piercy, F. P. (2005). Infidelity: An overview. *Journal of Couple & Relationship Therapy, 4*(2/3), 5–16.

Humphrey, F. G. (1987). Treating extramarital sexual relationships in sex and couples therapy. In Weeks, G. R., & Hof, L. (Eds.), *Integrating sex and marital therapy.* New York, NY: Brunner/Mazel.

Kaslow, F., & Hammerschmidt, H. (1992). Long-term "good marriages": The seemingly essential ingredients. *Journal of Couples Therapy, 3,* 15–38.

Kessel, D. E., Moon, J. H., & Atkins, D. C. (2007). Research on couple therapy for infidelity: What do we know about helping couples when there has been an affair? In Peluso, P. R. (Ed.), *Infidelity: A practitioner's guide to working with couples in crisis* (pp. 55–70). New York, NY: Routledge.

Linquist, L., & Negy, C. (2005). Maximizing the experiences of an extra-relational affair: An unconventional approach to a common social convention. *Journal of Clinical Psychology/In Session, 61*(11), 1421–1428. DOI: 10.1002/jclp.20191

Lusterman, D. D. (2005). Helping children and adults cope with parental infidelity. *Journal of Clinical Psychology/In Session, 61*(11), 1439–1451.

Nelson, T., Piercy, F. P., & Sprenkle, D. H. (2005). Internet infidelity: A multiphase Delphi study. *Journal of Couple & Relationship Therapy 4*(2/3), 173–194.

Pittman, F. S., & Wagers, T. P. (2005). Teaching infidelity. *Journal of Clinical Psychology, 61,* 1407–1419.

Rasmussen, P. R., & Kilborne, K. J. (2007). Sex in intimate relationships: Variations and challenges. In Peluso, P. R. (Ed.), *Infidelity: A practitioner's guide to working with couples in crisis* (pp. 11–30). New York, NY: Routledge.

Spring, J. A. (1996). *After the affair: Healing the pain and rebuilding the trust when a partner as been unfaithful.* New York, NY: HarperCollins.

Vaughan, P. (2003). *The monogamy myth.* New York, NY: Newmarket Press.

Whisman, M. A., Dixon, A. E., & Johnson, B. (1997). Therapists' perspectives of couple problems and treatment issues in couple therapy. *Journal of Family Psychology, 11,* 361–366.

Wubbolding, R. E. (1988). *Using reality therapy.* New York, NY: Harper & Row.

5

RECOVERING FROM SUBSTANCE MISUSE

Thomas K. Burdenski, Jr.

INTRODUCTION

Choice theory and reality therapy have been widely used in conjunction with the 12 steps of Alcoholics Anonymous and Al-Anon for recovery from alcoholism and co-dependency since the 1970s (Wubbolding & Brickell, 1999). The application of choice theory/reality therapy to couples has been emphasized in recent works by both Glasser (1998, 2000; Glasser & Glasser, 2000, 2007) and Wubbolding (Christensen & Gray, 2002; Wubbolding 2000a, 2000b, 2011; Wubbolding & Brickell, 1999). Behavioral couples therapy (BCT; O'Farrell & Fals-Stewart, 2006), an approach that focuses on both substance use and repairing relationships, has the best research support for helping couples face the challenges of recovery together. BCT is also highly compatible with choice theory/reality therapy principles as well as 12-step group participation.

Recent research has indicated that including partners of substance "misusers" (Van Wormer & Davis, 2008) in conjoint treatment can produce more abstinence, happier relationships, fewer separations, lower risk of divorce, substantially reduced domestic violence, fewer emotional problems in children, and lower treatment costs than individual treatment for substance misuse (O'Farrell, 1993; O'Farrell & Cutter, 1984; O'Farrell & Fals-Stewart, 2000). In BCT, the major goals are to alter interactional patterns that maintain problem drinking and to improve the relationship so that turning to alcohol becomes less attractive as a means of coping with stress and conflict (Thomas & Corcoran, 2001).

CHOICE THEORY AND REALITY THERAPY

The choice theory/reality therapy approach (Wubbolding & Brickell, 1999) is highly applicable to addictions counseling because of the necessity in early addictions counseling to focus on current (total) behaviors, the risks of continuing to use drugs or drink alcohol, and taking personal responsibility for one's behavior. The WDEP (wants, doing/direction, self-evaluation, and planning) delivery system is highly compatible with the counseling approaches used in most in-patient and outpatient recovery programs, and reality therapy, as a counseling approach, can be effectively applied in all stages of recovery (Gorkski, 1989). In a $26 million study of nearly 2,000 alcohol-dependent individuals (Project MATCH, 1997), outpatients who received the individually administered 12-step treatment did as well as participants receiving individually administered cognitive-behavioral or motivational enhancement treatment and they were *more likely* to remain completely abstinent in the year following treatment (Van Wormer & Davis, 2008). Attendance at AA meetings was encouraged more by counselors applying 12-step facilitation than counselors applying cognitive-behavioral or motivational enhancement treatments.

Wubbolding and Brickell (1999) concluded:

> The continuity of treatment and undoubted sense of *belonging, connectedness,* and *support* provided by "Twelve-Step Facilitation," plus its emphasis on treating the "spiritual disease" within (a more holistic approach overall), will prove to be the most successful of the three modalities *in the longer term.* (p. 101)

Wubbolding and Brickell (1999) asserted that reality therapy will usually be most effective when combined with a long-term program that provides ongoing support and meets the recovering alcoholic's (or substance misuser's) need for *love and belonging.* They note that the great strength of 12-step recovery groups from a choice theory perspective is that recovering alcoholics and drug addicts can rebuild their lives by learning to meet their five basic needs in healthy and satisfying ways and to connect (or reconnect) with others who can provide caring, acceptance, trust, and support, as well as serve as models for living who share their experience, strength, and hope.

According to Wubbolding (1999), couples have a successful intimate relationship if there is overlap in terms of specific desires, wants, and goals related to the sources of human motivation: the five basic needs. The five needs provide the fuel and the energy that draw couples closer together in compatible relationships and they can also drive couples apart if these needs conflict or are not negotiated. Couples need to discuss the intensity of the five basic needs in order to maximize pleasure and minimize pain in their relationships and they also need to build quality time as the foundation for closeness and intimacy. Wubbolding and Brickell (1999) describe four levels of interaction in couples that progress from the least demanding and least intimate (level D) to the most demanding and most intimate (level A). When levels D and C are being met, each partner develops a storehouse of positive memories and mutually positive perceptions that allow for altruistic concern that the other's quality world is realized. After individual needs are articulated and addressed in level B, the couple is ready to move to level A, where the focus is on decision making, problem solving, and compromise.

BEHAVIORAL COUPLES THERAPY

Similar to the WDEP process of choice theory and reality therapy, behavioral couples therapy (O'Farrell & Fals-Stewart, 2006) involves guiding the couple through progressive phases: a "recovery contract" and four other strategies to promote abstinence; four strategies to improve the couple's relationship; and, finally, taking steps to continue recovery after counseling ends. In the WDEP system of reality therapy, the initial step is to identify what the clients want to see happen. The focus is on the positive, desired outcome (the quality world picture of abstinence) rather than the problematic behavior of substance misuse. In the BCT method, the process begins with creating a recovery contract in which both persons support the substance misuser's abstinence, which is what they want or their goal. The couple makes a joint commitment to create a happier relationship by rewarding behaviors in one another that support abstinence and long-term recovery and learning communication and conflict-resolution skills to increase commitment and positive feelings in the relationship.

The second step of the WDEP process is to identify what current behavior is being used in an effort to get what the clients want. The BCT recovery contract supports abstinence by specifying which behaviors each party can decrease to increase trust and reduce conflict about substance misuse and to reward actions that support abstinence. The contract begins with a "trust discussion" that is continued daily by having the misuser share his or her intention not to misuse that day, in the tradition of AA's "one day at a time." This is followed by the spouse's expressed support for the other's abstinence and, lastly, the misuser thanking the spouse for his or her encouragement and support. Participation in self-help groups for both parties is encouraged and committed to on a "recovery calendar" provided by the counselor, and both partners commit to not discussing substance-related conflicts that can trigger relapse outside counseling sessions.

The recovery contract and calendar are important tools because the counselor reviews the calendar with the couple at the start of each session to see how well each partner has followed through on the commitments and to reinforce the couple's progress verbally. As in the self-evaluation component of the WDEP process, the counselor also asks the couple to review their recovery contract out loud to underscore its importance, identify any modifications that might be made, and give feedback on how each partner expresses his or her needs.

The final step of the WDEP system is creating a plan for future success. In addition to the recovery contract, the BCT plan includes four more tools that support abstinence and help deter relapse (O'Farrell & Fals-Stewart, 2006):

- "Reviewing substance use or urges to use" involves openly discussing situations, thoughts, and feelings that trigger use and any successful coping strategies that the substance misuser has engaged to resist these urges in the past. If a relapse occurs, it is addressed by the counselor immediately and treated as a learning experience.
- "Decreasing exposure to alcohol and drugs" involves discussing situations in which alcohol or drugs are used, how to avoid those situations whenever possible, and how to deal with situations that might arise in the future.
- "Addressing stressful life problems" focuses on assisting the couple to address financial, legal, and other problems that may increase relapse risk. This step also builds good will between the counselor and the client since overall welfare is addressed, not just stopping alcohol or drug use.
- "Decreasing partner behaviors that trigger or reward use" involves having the person in recovery identify the past behaviors practiced by the nondrinking spouse, such as lying to cover for the substance misuser or giving the misuser money to buy alcohol or drugs, that may have contributed to using or relapsing in the past.

Glasser (1998) noted that relationship problems are at the source of most long-term problems. The first step of the recovery contract focuses on building a better relationship while also identifying and sharing mutual wants and goals. Using the BCT model, after the couple has effectively implemented the recovery contract and is working together effectively to establish and maintain abstinence, the focus turns to "improving the couple's relationship." In many couples, the biggest emotional barrier to recovery for the misuser is guilt about past behavior and wanting to get more credit for current changes than he or she has received thus far. For the partner, there are often resentments about past behavior, fears of relapse, and distrust about the prospects for change.

After years of drinking or drug use, couples often become estranged and unable to address and resolve many of the small issues that healthy couples face and work through each day. There are often communication skills deficits due to not learning how to communicate and solve problems collaboratively. Focusing on the relationship is designed to help smooth the role transition period as the misusing partner takes on more responsibilities in the home and the nonmisusing partner begins to relinquish control as trust builds in the recovering partner.

The "improving the couple's relationship" stage includes both increasing positive activities and improving communication. There are two major activities that promote positive contact. The first is "catch your partner doing something nice" (O'Farrell, 1993), in which each partner keeps track each day of at least one nice thing that the other did. In a "caring day," partners take turns planning ahead for a special activity to demonstrate how much each person cares for the other. In "shared rewarding activities" (Noel & McGrady, 1993), couples plan and carry out mutually enjoyable activities. Planning and participating in shared rewarding activities is important because research has demonstrated a connection between couples reconnecting and enjoying one another's company with higher abstinence and lower rates of relapse (Moos, Finney, & Cronkite, 1990).

The couple's communication skills are improved by teaching the couple communication skills, such as the reflective listening (paraphrasing, empathizing, and validating), conflict resolution skills, learning to negotiate agreements for desired changes, and developing problem-solving skills. All of these are designed to help the couple to address their stressors in a constructive fashion and learn to communicate concerns without blaming or resorting to other behaviors that could provoke a relapse (O'Farrell & Fals-Stewart, 2006). Similarly, Glasser has emphasized in recent years (1998, 2000, 2002; Glasser & Glasser, 2007) practicing relationship-building habits that he calls "the seven caring habits" (supporting, encouraging, listening, accepting, trusting, respecting, and negotiating differences) instead of indulging in relationship-destroying habits.

The final stage of BCT treatment is the "continuing recovery" stage. As with all counseling, there is a tendency when working with couples for backsliding to occur as they move away from a structured weekly counseling meeting to terminating counseling and applying their newly learned skills on their own. Near the end of the counseling period, the counselor helps the couple make a "continuing recovery plan" that details which aspects of the treatment (recovery plan, activities that support abstinence, increasing positive time together, and improving communication skills) they wish to retain after the formal counseling period ends. They are also asked to develop an action plan that clearly spells out what they will do to prevent or minimize relapse. If possible, couple checkup visits are arranged every few months to encourage continued progress.

CASE STUDY: DON AND CARMEN

Presenting Problem

While Don is progressing steadily in his recovery from long-term alcohol dependence, he and his wife, Carmen, are having arguments daily over his role as parent and decision maker in the family. Don gradually relinquished more and more authority in the family as his drinking became worse, his involvement in the family diminished, and Carmen took over as the head of the household and as the primary parent to their two children, now both in high school. Now that he is sober, Don is anxious to regain some of his parental authority but Carmen is reluctant to trust him and give up any of her authority because the family has been running smoothly (with the exception of Don's problematic drinking) and she is afraid that he might relapse.

The arguments are beginning to upset their children and they are both experiencing deterioration in their school performance. Don and Carmen are beginning to think that divorce may be the only solution to their conflicts and Don is beginning to question whether all of his efforts to stay sober are worthwhile. Don and Carmen met with a couples' counselor to discuss the possibilities of working things out and saving the marriage. They have agreed to attend 16 weeks of choice theory/reality therapy couples counseling supplemented by behavioral couples therapy interventions.

Background Information

Don is a 48-year-old marketing representative in the oil industry who has gradually developed alcohol dependence as a result of gradual increases in his alcohol consumption over a period of 30 years. He has been married to Carmen, 45, for 21 years and they have two teenage children living at home, both attending high school. Don stopped drinking 2 months ago when he was arrested for driving under the influence (DUI) on his way home from entertaining a longtime client over dinner and several rounds of drinks. Since completing a 30-day detoxification program 1 month ago, Don has been sober and has not missed any work days. He sees his probation officer regularly and has found an Alcoholics Anonymous fellowship with meetings that he really likes attending. He often attends three or more meetings a week, but he has not yet found a sponsor.

Assessment

Don and Carmen are undergoing a fairly typical role adjustment that occurs when the newly sober spouse strives to restore his or her place in the family. This represents a threat to the equilibrium of the family system that was established to fill the void left by Don's lack of involvement in the family. Carmen and Don's two children gradually learned to adjust to his shrinking role as parent and family decision maker and both children became very close to Carmen as they grew older. To Carmen, the status quo with her having the majority of parental and household authority "is working just fine" and she is reluctant to "fix something that is not broken."

Case Conceptualization

In many ways, Don and Carmen's struggle with realigning family roles after the drinking partner has obtained sobriety is characteristic of the conflicts and adjustments that families go through following sobriety. In AA, there is a tongue-in-cheek slogan that members share with the newly sober: "The only thing that you have to change is everything." The transition to sobriety calls for all family members to make role adjustments and to work cooperatively to help the recovering family member restore his or her place in the family. The best treatment strategy is to get both partners working jointly on the many challenges faced by sobriety and to begin to repair rifts in the relationship caused by many years of problematic drinking.

Treatment Goals

Glasser (1998) asserts that most long-term psychological problems are due to the lack of a need-fulfilling relationship with another person. The goal of choice theory/reality therapy (CT/RT) couples counseling is first to determine if both partners want to improve the relationship. If there is a commitment to strengthening the relationship, the focus shifts to having the couple work together jointly and cooperatively to assist the person in recovery with establishing and maintaining abstinence. Once abstinence is maintained and the couple/family has had a chance to stabilize, the CT/RT couples counselor can help the couple clarify both relationship strengths and problem behaviors and encourage each partner to identify his or her part in problematic interactions. This is followed by encouraging the couple to engage in new behaviors using a SAMIC (simple, attainable, measurable, immediate, consistent; Wubbolding, 2000a, 2011) plan to improve the relationship in small ways as quickly as possible.

Improving the relationship often means addressing the "quality time" spent with one another—time spent building intimacy and sharing activities that are mutually need fulfilling. This may involve returning to activities that brought the couple together in the early stages of their relationship or finding new ways to connect that meet one another's five basic needs. Planning together how those needs will be optimized constitutes the treatment goals.

Summary of Counseling Process Supplementing RT/CT With BCT

The 16-session treatment followed the BCT emphasis on establishing and maintaining sobriety—first, by focusing on the recovery contract, then by helping the couple improve their relationship by increasing positive activities and improving communication skills, and by closing with a continuing recovery plan to reduce the risk of backsliding and relapse.

Session 1 In the first session, the couple expressed a commitment to keeping the marriage intact and Don and Carmen took active steps to support Don's continued sobriety. They committed to 16 sessions of CT/RT counseling and agreed to have a "trust discussion" each morning over coffee when they would review several agreements. Don committed to attend at least three AA meetings each week, to take Antabuse (disulfiram) in front of Carmen each morning, and to keep his prescription filled. Antabuse is a drug that produces extreme nausea when the person taking it drinks alcohol, and including Antabuse in the trust discussion has improved pharmacotherapy compliance with alcoholic patients (Azrin, Sisson, Meyers, & Godley, 1982; Chick et al., 1992; O'Farrell, Cutter, Choquette, Floyd, & Bayog, 1992). Carmen agreed to attend three Al-Anon meetings each week and committed to not discussing Don's past drinking or her fears about Don's future drinking. The couple completed the recovery contract, agreed to keep the recovery contract calendar up to date, and decided to keep it posted on the refrigerator for easy reference.

In the recovery contract, Don agreed to reaffirm his intention to Carmen to stay sober that day; take the Antabuse; thank Carmen for listening and observing him take the Antabuse; remind Carmen not to mention his past drinking or her fears about future drinking (if necessary); acknowledge to Carmen her support for him and exercising restraint in not being accusatory; refill his Antabuse prescription before it runs out; and attend three AA meetings that week. Carmen agreed to record Don's intention to stay sober that day on the calendar; observe Don taking the Antabuse and record it on the calendar; thank Don for reaffirming his intention to stay sober and for taking the Antabuse; and not mention Don's past drinking or her fears about future drinking.

The counselor asked the couple to practice their trust discussion in front of him to highlight its importance and to observe how they did it. While practicing, Don expressed anger over having to be "checked up on" by Carmen. The counselor clarified to both that the purpose of the contract is to rebuild trust, rather than to coerce Don into not drinking. With that reframing, Don calmed down and developed a more positive attitude about the morning ritual. When it came time for Don to practice thanking Carmen for her encouragement and support, Carmen let Don know that his words of appreciation were really meaningful to her because "you got in trouble with drinking and I am working really hard to help you stay sober." They agreed to go to AA and Al-Anon meetings scheduled at the same time so that they could use the drive to and from meetings as a chance to connect with one another and share quality time.

Sessions 2–5   In Sessions 2–5, Don and Carmen talked about situations, thoughts, or feelings that had triggered past use, ways to resist urges, ways to decrease Don's exposure to alcohol, and how to handle situations that involve others drinking alcohol in Don's presence. Don shared feeling criticized by Carmen about his parenting and that having her "tell me how to be a dad" was the biggest trigger to past drinking. He said, "I get so frustrated and angry … I know I can't hit her, so I drink at her." Carmen said, "It's a catch-22: He hasn't been the most involved dad and if I say anything then I'm harping on him." Don also shared that it had been really hard for him to "watch her drink a nice bottle of wine." Carmen quickly agreed not to drink in front of Don, but Don then expressed concern about having wine in the house at all. "When I open the refrigerator and see a wine bottle, it's a big temptation," he said. The couple eventually agreed that Carmen would limit her drinking to outside the home on Friday evening with friends.

Don also shared that he was unsure of how to handle not drinking around clients over lunch, when playing golf with clients, and after a particularly trying day at work, especially after being criticized by clients or his supervisor. "There's no relief," he said. He shared that Antabuse is an effective deterrent to relapsing and that going through treatment helped him feel not so alone with battling his disease. He added that he was "learning to pray" when he was distressed and that he had begun to call other men in AA when he felt tempted to drink, although he had not yet found a sponsor. Don said that he appreciated Carmen's genuine desire and support for his sobriety and that he felt "motivated to not let her down."

Don was very happy with the support he was getting from his AA fellowship and he shared that he felt completely supported when he recently took a risk and reached out to an AA friend after experiencing an urge to drink. To build goodwill and take some of the focus off alcohol, the counselor addressed other legal, financial, and family problems that might increase relapse risk. The couple talked about lawyer fees, court costs, and fees to the probation department resulting from Don's DUI. Don added that he was expecting a promotion, and both he and Carmen agreed that the extra income will relieve the temporary financial stress they were currently experiencing. The counselor encouraged Don to consider asking his AA friend to be his sponsor.

Finally, the couple discussed Carmen's past behaviors that may have triggered or rewarded Don's alcohol use and ways to prevent those behaviors from recurring. Don discussed feeling self-conscious around clients with whom he used to drink, but he eventually concluded that "I must be much more concerned than they are because only one person has even brought it up." Carmen added that another struggle was that most of their friends who are couples drink and that she was confused about what to do with their social life because they had stopped getting together with their wide circle of friends. Don and Carmen then had a constructive conversation about with whom Don was comfortable sharing his sobriety and commitment to abstinence, and he agreed to let Carmen share that information with Connie and Bob, two of their closest friends.

Don and Carmen then briefly touched on their discomfort with sexual intimacy after Don's sobriety. Carmen added that "everything is different about Don since he stopped drinking" and that when she began couples counseling she was unaware of how much her life also needed to change—not just Don's. "I didn't put it together that I was going to have change everything too," she said. The counselor promised to come back to the intimacy issue later in the course of couples counseling after the couple had a chance to stabilize further and learn healthy communication skills.

Sessions 6–8 After Session 5, the counselor determined that the abstinence period was becoming firmly established because Don and Carmen were coming to their weekly sessions, Don remained sober, and they were reviewing their recovery contract with one another almost every morning. In Sessions 6–8, Don and Carmen were encouraged to address stressful life problems by sharing the underlying feelings that seemed to fuel their frequent arguments. Don shared his guilt about his past drinking and the time that he had lost with Carmen and their two children. He also expressed frustration that Carmen did not seem to appreciate all of the effort he was spending on his abstinence and recovery. Carmen shared her resentments about Don's past drinking and her fears that he might relapse. The counselor addressed the many role adjustments that Don, Carmen, and their two children had to make if they were to bring Don successfully "back into the family." Don expressed his frustration over Carmen's reluctance to consult him on decisions about the household or parenting their teenage children. Carmen brought out her fears that "Don's abstinence will not last and he will cause more problems for the family."

At the end of this session, the counselor encouraged Don to discuss any of Carmen's past behavior that might have triggered or rewarded alcohol use. Don shared that Carmen lied to cover for him on days that he was hung over and could not make it to the kids' sporting events or other family events. The counselor encouraged the couple to hold off on expressing their angry and resentful feelings further until they learned more functional communication skills, planned for later in the course of counseling. Don added that "from time to time we both avoid talking about things we don't want to face." Carmen agreed, saying, "For a long time when I was angry; I didn't say anything because I didn't want there to be a problem."

Sessions 9 and 10 After Don and Carmen aired many of the painful feelings that they were holding inside in Sessions 6–8, the counselor shifted to improving the couple's relationship by increasing positive interactions by assigning the "Catch Your Partner Doing Something Nice" activity. The counselor asked each partner to record one or more pleasing behaviors initiated by the other each day by filling out the weekly log sheet. Don and Carmen were then coached on how to express gratitude for one another's pleasing behaviors by emphasizing good eye contact, a smile, sincerity, and a positive tone of voice. For example, Carmen told Don, "I liked it when you were patient with me when I came home tired and irritable from work two nights ago—your upbeat sense of humor made it hard for me stay grumpy."

Don and Carmen were then assigned the task of sharing with one another, in a 5-minute conversation, one pleasing behavior that they appreciated each day. They decided to enact this ritual right before bedtime each night. In the "Caring Day" assignment, Don and Carmen picked a day during the coming week to perform loving acts toward one another, which encouraged both of them to take risks and initiate loving kindness without waiting for the other party to make the first move. Don was hesitant to apply himself to this activity, so the counselor reminded him that he had agreed at the outset of counseling to act differently and then assess the impact of his loving behavior, rather than waiting until he was ready to show his love. Don "came around" when the counselor explained that the "Caring Day" did not have to involve expensive gifts, but rather the willingness to extend himself to do and say things that helped Carmen to feel loved.

In the "Planning Shared Rewarding Activities" exercise, the counselor began by discussing all the ways that planning and enjoying quality time can be sabotaged (waiting until the last minute so that proper planning cannot be done, getting side-tracked, not keeping to the time schedule, etc.). The counselor then asked Don and Carmen to make separate lists of fun activities that they enjoyed and that they could enjoy with one another. Both Don and Carmen identified taking walks together, hiking on the weekends, and sharing coffee in the morning as rewarding shared activities (activities common to each of their quality worlds). They then collaboratively scheduled these activities into their recovery calendar.

Sessions 11–14   In the next phase of counseling, the counselor began training in communication skills by asking Don and Carmen to share with one another positive changes that had occurred in their marriage and in the family since the beginning of counseling. They agreed that Don's recent promotion at work and that having the entire family sit and eat dinner together were major improvements and that the family was beginning to relax and enjoy one another more. They then moved on to discussing more emotionally charged topics, like their lingering disagreement about sharing authority for family decision making and Carmen not including Don in family matters and disciplinary decisions that affected their two children.

As is typical of most conflicted couples, Don and Carmen had to learn to slow their discussions way down when discussing emotionally charged topics to facilitate staying on track and to avoid the temptation to bring up the past or reacting defensively to one another. Because basic listening and speaking skills were now in place, the counselor added the additional task of accurately summarizing both the content and the feelings shared by the partner before the partner was allowed to share his or her point of view. For example, when discussing the parenting issue with Carmen, Don summarized what Carmen shared by saying:

> What I heard you say is that you want me to be more involved with parenting and family decisions, but you are afraid that if we change things up right now with the kids nearly finished with high school, things might get worse in the family and, moreover, if you let me share the parenting role more fully, you may not be as close to our kids when they go off to college. Did I hear that right?

Only when Don heard Carmen fully was he given permission to share his thoughts and feelings on this topic. In the beginning, both Don and Carmen were under the misimpression that understanding the other's point of view and sharing it out loud meant agreeing with the partner's point of view. When the counselor dispelled this notion, they both improved dramatically in their ability to hear and paraphrase the other's thoughts and feelings.

In Session 12, Don and Carmen were coached more thoroughly on how to express both negative and positive feelings more directly by taking full responsibility for feelings, not blaming the other for negative feelings, and using "I" statements instead of "you" statements. The counselor then modeled how to share positive and negative feelings effectively and asked Don and Carmen to practice sharing positive and negative feelings in the session. They were then instructed to schedule three 10- to 15-minute communication practice sessions before the next counseling session to fine-tune their skills, beginning with easier topics before progressing to more difficult topics.

In Session 13, Don and Carmen were taught how to listen attentively without any interruptions for longer and longer periods of time before summarizing the speaker's message and feelings about the topic. The counselor began by having the couple practice sharing their thoughts and feelings about receiving caring behaviors for 2 minutes each. When Don and Carmen demonstrated proficiency with listening for short periods of time, the counselor shifted to longer periods of time (5–10 minutes) in which the couple talked about problems or concerns in the relationship. As homework for week 13, Don and Carmen were assigned three 10- to 15-minute practice sessions to deepen their skill at listening accurately and with empathy to the partner sharing emotionally charged topics.

In Session 14, Don and Carmen were taught how to make specific positive requests of one another and to negotiate and compromise with one another so that some of the long-standing issues could be resolved. The counselor demonstrated how to break out of the habit of complaining to the partner or using browbeating, coercion, or anger to try to force the partner to change. The counselor then coached the couple to convert the positive request into a mutually agreed upon action plan for the coming week. Carmen requested that Don spend 10 minutes each evening listening intently to her discuss her day without interruption, paying special attention to her feelings, paraphrasing occasionally, and taking pains not to attempt to solve Carmen's problem or change the subject. Don requested that Carmen wait to talk over family and household decisions before making plans for the family or responding to their two children when they approached her for advice or to discuss a problem so that Don could become more involved in the parenting role.

Session 15: Termination In the final session, the counselor reviewed which aspects of the recovery plan were most useful to the couple and they negotiated a "continuing recovery plan" that they agreed to follow on their own. The counselor shared that there is a risk of backsliding after the couple's counseling ends because of the structure and accountability built into the weekly counseling sessions. The counselor also asked the couple to discuss situations that presented a high risk for relapse after terminating counseling. Dan and Carmen both expressed concern that they might slip back into their old pattern of avoiding conflict and sitting on negative feelings (like anger, hurt, and disappointment) by "making nice" and "pretending that everything is just fine." The counselor asked Don and Carmen to make separate lists on what topics they were most likely to avoid honestly sharing with one another and then he coached the couple on how to handle these. The counselor emphasized the many strengths and resources (especially their newly developed communication skills) that each partner could turn to in times of high stress and the couple then rehearsed coping strategies for such events.

The counselor strongly emphasized the critical importance of early intervention at the beginning stage of a relapse as opposed to waiting until the alcohol reached dangerous levels with serious damage to couple and family relationships. Finally, Don and Carmen were encouraged to include continued contact with the counselor for 3–5 years after sobriety had been established in their long-term recovery plan. This could include telephone checkups or "fine-tuning" sessions to support their ongoing recovery because substance misuse is a chronic health problem that requires active, aggressive, ongoing monitoring to prevent or re-treat relapses for 5 years after sobriety is established. The follow-up sessions would also provide the couple the opportunity to get assistance with problems that were not addressed in counseling or did not arise during treatment.

When the counselor asked the couple which choice theory, reality therapy, or behavioral couples therapy techniques they benefited from the most, Don said that he really liked the communication exercise in which Carmen had to keep repeating what he said until she was able to share his point of view to his satisfaction. "I felt like she actually heard me," he said. "That's one of the things that really helped me to work things out with Carmen better." Both Don and Carmen commented on how much they valued the "day of caring." Carmen said, "I noticed that there were a lot of good things happening between us that I wasn't recognizing before we did the caring day because I was focusing on all of the problems in the relationship—not what was going well."

DISCUSSION

When Don and Carmen had completed 16 sessions of couples counseling, there were major improvements in the family's adjustment to Don's sobriety and enlarged role within the family. Couples therapy for couples in recovery takes time. Brown and Lewis (2002) state:

> As long as the abstinence is seen as the end of the problem, which is convenient for all, rather than the beginning of a new growth process that can involve all, the myth will continue: Recovery should bring a reversal of trauma and much improved family function and relationship in a very brief period of time. All of these changes do occur, but not quickly. There is more disruption and turmoil that comes first—what we call the trauma of recovery. (p. 7)

As is often the case, Don, Carmen, and their two teenage children all underestimated how much Don's sobriety would disrupt the family equilibrium. There was so much excitement initially about his sobriety that no one really thought through how difficult it would be for every member of the family to make a major role adjustment. Carmen had the most difficult time of all because she had liked having a lot of power as head of the family and she took a lot of pride in how well her children were doing before Don entered treatment and became sober. Initially, it appeared to everyone that the family was doing worse as a result of Don's recovery because Don and Carmen were in a high degree of conflict and both were seriously considering a divorce, which affected their children to the point that they were both performing poorly at school.

Couples like Don and Carmen, who are trying to rebuild their lives and families, often have an uphill struggle because many couples divorce during the first year of sobriety (O'Farrell & Fals-Stewart, 2006). Joan Jackson (1954, 1956) described the stages that families go through as the drinking spouse becomes sober. Jackson found that women took on more and more of the family roles as their husbands' alcoholism worsened, much like Carmen did. As the alcoholism progressed, the nonalcoholic partner typically took on more of the parenting, financial decision making, and financial support of the family because the alcoholic spouse could not be counted upon to carry out these roles. Similarly, Brown and Lewis (2002) concluded that "abstinence marks the beginning of a new developmental process that has a profound, complicated impact on the whole family" (p. 6). Substance misuse and alcohol dependency are a chronic health problem that requires active and aggressive treatment and active and aggressive ongoing monitoring after treatment ends to prevent or quickly treat relapses for at least 5 years after the initial period of recovery becomes stabilized (O'Farrell & Fals-Stewart, 2006).

Supplementing choice theory/reality therapy for couples with BCT is important because marital and couple relationships where one partner misuses substances are typically marked by conflict and dissatisfaction. BCT attempts to increase commitment and positive feelings by improving communication and conflict-resolution skills. Such improvements can increase motivation to continue treatment and decrease the likelihood of divorce after abstinence has been achieved. BCT has been shown to improve couples' participation in substance misuse treatment and treatment outcomes (Steinglass & Kutch, 2004), as well as improving relations between partners (Jacobson et al., 1984).

O'Farrell and Fals-Stewart (2006) describe the clinical stance of BCT as follows:

BCT encourages a good-faith-individual-responsibility approach in which each member of the couple freely chooses to make needed changes in his or her behavior independent of whether or not the partner makes needed changes in his or her behavior. ... It is only human nature that each member of the couple will want the other to change first and also will want to stop his or her own efforts if the partner does not seem to be making a serious effort to change. (pp. 3–4)

In summary, choice theory/reality therapy and behavioral couples therapy make excellent partners for counselors assisting couples who are facing the challenge of overcoming alcohol or drug dependence. The philosophies of the two approaches are highly compatible and the thoroughly researched techniques and approaches to teaching communication skills utilized in BCT have a strong record for clinical effectiveness. See O'Farrell and Fals-Stewart (2006) for an in-depth explanation of how to put their 16-session BCT format into clinical practice with couples.

REFERENCES

Azrin, N. H., Sisson, R. W., Meyers, R., & Godley, M. (1982). Alcoholism treatment by disulfiram and community reinforcement therapy. *Journal of Behavior Therapy and Experimental Psychiatry, 13,* 105–112. doi:10.1016/0005-7916(82)90050-7

Brown, S., & Lewis, V. (2002). *The alcoholic family in recovery: A developmental model.* New York, NY: Guilford Press.

Chick, J., Gough, K., Falkowski, W., Kershaw, P. Hore, B., Mehta, B., Ritson, B.,... Torley, D. (1992). Disulfiram treatment of alcoholism. *British Journal of Psychiatry, 161,* 84–89. doi:10.1192/bjp.161.1.84

Christensen, T. M., & Gray, N. D. (2002). The application of reality therapy and choice theory in relationship counseling, an interview with Robert Wubbolding. *The Family Journal, (10)*2, 244–248. doi:10.1177/1066480702102020

Glasser, W. (1998). *Choice theory: A new psychology of personal freedom.* New York, NY: HarperCollins.

Glasser, W. (2000). *Counseling with choice theory: The new reality therapy.* New York, NY: HarperCollins.

Glasser, W. (2002). *Unhappy teenagers: A way for parents and teachers to reach them.* New York, NY: HarperCollins.

Glasser, W., & Glasser, C. (2000). *Getting together and staying together: Solving the mystery of marriage.* New York, NY: HarperCollins.

Glasser, W., & Glasser, C. (2007). *Eight lessons for a happier marriage.* New York, NY: HarperCollins.

Gorkski, T. T. (1989). *Passages through recovery: An action plan for preventing relapse.* New York, NY: Harper & Row.

Jackson, J. K. (1954). The adjustment of the family to the crisis of alcoholism. *Quarterly Journal of Studies on Alcohol, 15,* 562–586.

Jackson, J. K. (1956). The adjustment of the family to alcoholism. *Marriage and Family Living, 18,* 361–369.

Jacobson, N. S., Follette, W. C., Revenstorf, D., Baucom, D. H., Hahlweg, K., & Margolin, G. (1984). Variability in outcome and clinical significance of behavioral marital therapy: A reanalysis of outcome data. *Journal of Consulting and Clinical Psychology, 52*(4), 497–504. doi:10.1037/0022-006X.52.4.497

Moos, R. H., Finney, J. W., & Cronkite, R. C. (1990). *Alcoholism treatment, context, process and outcome.* New York, NY: Oxford University Press.

Noel, N. E., & McCrady, B. S. (1993). Alcohol-focused spouse involvement with behavioral marital therapy. In T. J. O'Farrell (Ed.), *Treating alcohol problems: Marital and family interventions* (pp. 210–235). New York, NY: Guilford Press.

O'Farrell, T. J. (1993). A behavioral marital therapy couples group program for alcoholics and their spouses. In T. J. O'Farrell (Ed.), *Treating alcohol problems: Marital and family interventions* (pp. 305–326). New York, NY: Guilford Press.

O'Farrell, T. J., & Cutter, H. S. G. (1984). Behavioral marital therapy for alcoholics: Clinical procedures from a treatment outcome study in progress. *American Journal of Family Therapy, 12,* 33–46. doi:10.1080/01926188408250183

O'Farrell, T. J., Cutter, H. S. G., Choquette, K. A., Floyd, F. J., & Bayog, R. D. (1992). Behavioral marital therapy for male alcoholics: Marital and drinking adjustment during the two years after treatment. *Behavior Therapy, 23,* 529–549.

O'Farrell, T. J., & Fals-Stewart, W. (2000). Behavioral couples therapy for alcoholism and drug abuse. *Journal of Substance Treatment, 18*(1), 51–54. doi:10.1016/S0740-5472(99)00026-4

O'Farrell, T. J., & Fals-Stewart, W. (2006). *Behavioral couples therapy for alcoholism and drug abuse.* New York, NY: Guilford Press.

Project MATCH Research Group (1997). Matching alcoholism treatments to client heterogeneity: Project MATCH posttreatment drinking outcomes. *Journal of Studies on Alcohol, 58*(1), 7–29. doi:10.1111/j.1530-0277.1998. tb03912.x

Steinglass, P., & Kutch, S. (2004). Family therapy: Alcohol. In M. Galanter & H. Kleber (Eds.), *The American Psychiatric Publishing textbook of substance abuse treatment* (3rd ed., pp. 405–415). Arlington, VA: American Psychiatric Publishing, Inc.

Thomas, C., & Corcoran, J. (2001). Empirically based marital and family interventions for alcohol abuse: A review. *Research on Social Work, 11*(5), 549–575. doi:10.1177/104973150101100502

Van Wormer, K., & Davis, D. R. (2008). *Addictions treatment: A strengths perspective* (2nd ed.). Stamford, CT: Brooks/Cole Cengage Learning.

Wubbolding, R. E. (1999). Creating intimacy through reality therapy. In J. Carlson & L. Sperry (Eds.), *The intimate couple* (pp. 227–246). Philadelphia, PA: Brunner/Mazel.

Wubbolding, R. E. (2000a). *Reality therapy for the 21st century.* Philadelphia, PA: Brunner/Mazel.

Wubbolding, R. E. (2000b). Brief reality therapy. In J. Carlson & L. Sperry (Eds.), *Brief therapy with individuals & couples* (pp. 264–285). Phoenix, AZ: Zeig, Tucker & Theisen.

Wubbolding, R. E. (2011). *Reality therapy: Theories of psychotherapy series.* Washington, DC: American Psychological Association.

Wubbolding, R. E., & Brickell, J. (1999). *Counseling with reality therapy.* Brackley, UK: Speechmark Publications LTD.

6

COUNSELING MILITARY COUPLES

Janet Fain Morgan

INTRODUCTION

Military couples face unique relationship challenges. When trials arise in this relationship, understanding the characteristics of the military society enhances appropriate treatment interventions and, hopefully, successful outcomes. Counseling military couples involves viewing the military culture as a distinct segment of society, with its own set of values, beliefs, and norms. It involves a plethora of acronyms, a complex chain of command, and an intricate rank structure for each branch of the armed forces. Not only are there independent cultures within each branch of the service, but these cultures also are divided into rank structures of officers and enlisted soldiers. The demographics of the military culture expand with each branch of service: Army, Navy, Marines, Air Force, and Coast Guard.

The core values of the Army embrace terms such as *loyalty, courage, commitment, integrity, honor,* and *selfless service.* Martin and McClure (2000) posit that

> Civilian human service providers will play an increasingly important role in the future delivery of broadly defined health and social services to military members and their families. … In most cases these services will come from the local community and from professionals who typically will not have personal experience in the

military. Service providers may have very little professional preparation for serving military populations. (pp. 20–21)

Counselors and providers that educate themselves in military culture and, in particular, the warrior culture and deployment science may enhance the therapeutic relationship and foster a positive counseling environment.

According to the National Healthy Marriage Resource Center (Hull, 2006), the active duty enlisted workforce has a mean age of 27 years. For officers, the mean age is 34 years. Both enlisted service members and officers comprise a much younger mean age when compared to their civilian working counterparts. Recruits that enlisted during this recent wartime period are better educated, wealthier, and have no race or ethnicity over-representation. This middle-class representation of recruits is predominantly male. They often have fairly clear and serious priorities of duty and honor and disciplined and focused determination to accomplish their goals. They are a younger workforce, and a more concise demographic portrait of active duty soldiers indicates that approximately half are married, a higher percentage when compared to their civilian peers (Kane, 2006).

STRESSORS FACING MILITARY FAMILIES AND COUPLES

Some major stressors that can influence military families are frequent moves, separations, job changes, foreign residence, foreign-born spouses, deployments to a war zone, and dual military roles. The aftereffects of combat experience and multiple deployments that a soldier experiences, known as combat operational stress (COS), may correlate to subsequent marital problems and an increase in the use of mental health services for both the active duty member and the spouse. Deployment of the spouse and length of deployment have been associated with an increase in mental health diagnoses among U.S. Army wives (Mansfield et al., 2010). One third of soldiers returning from deployment report using mental health services (Tanielian & Jaycox, 2008).

The percentage of deployed soldiers screening positive for mental health problems is currently lower than was seen in previous years. It is currently at 11.9% for a combined measure of acute stress, depression, or anxiety. However, the Mental Health Advisory Team of the U.S. Army Medical Command (2009) concluded that mental health issues, including marriage and relationship problems, generally increased with each subsequent month of the first two thirds of the deployment and then declined during the last third of the deployment. In addition, mental health and work-related problems reportedly increased exponentially with the number of deployments.

On the other hand, the Mental Health Advisory Team VI (MHAT VI) reported that the marital satisfaction rating declined significantly in the past 6 years but was critically lower for the E1 through E4 (mostly younger and inexperienced) ranks than for noncommissioned officer (NCO) ranks. Self-reporting results of divorce or separation intent was similar to earlier results at 16.5%, but the MHAT VI surveys also revealed that the low marital satisfaction rating was unrelated to number of deployments or dwell time.

SUPPORT SERVICES

In response to these conflicting data, the Department of Defense (DoD) has increased the number of supportive and preventive programs designed for military couples to strengthen their relationships and prepare for deployments to war zones. In addition to family readiness programs (FRG) that provide families with information about their deployed soldiers, the military has created preparation for deployment programs for families and postdeployment programs that provide information on how to prepare for deployment, how to cope during deployment, and how to integrate back into family life once the soldiers return home.

The Defense Centers of Excellence (DCoE, 2010b) states that its mission, "assesses, validates, oversees and facilitates prevention, resilience, identification, treatment, outreach, rehabilitation, and reintegration programs for psychological health and traumatic brain injury to ensure the Department of Defense meets the needs of the nation's military communities, warriors and families." In addition, the DoD offers a marriage and relationship enhancement guide (DCoE, 2010a) that assists marriage educators, chaplains, healthcare professionals, social service providers, health and wellness staff, and prevention and outreach workers with their work in promoting healthy couple relationships within military and veteran families. Primarily a compilation of resources, this guide also provides overviews of relationship enhancement education curricula, educational handouts, and information for couples who are looking for methods to enhance their marriage.

There is even a special Sesame Street workshop video (*When Parents Are Deployed*), available in English and Spanish, for children to learn age-appropriate aspects of deployment. The Army Behavioral Health Department, another division of the U.S. Army Medical Command, created a website (http://www.behavioralhealth.army.mil) dedicated to comprehensive mental and physical health of the soldier and family. The website offers training for licensed providers. To help reduce the effects that war has on individuals and families, there are dedicated training sites throughout the United States designed to train professionals.

CHOICE THEORY AND REALITY THERAPY

Choice theory and reality therapy, developed by Dr. William Glasser (1998), offer the military couple a therapeutic approach that not only emphasizes self-evaluation and personal responsibility—two areas embraced by the military culture—but also invites the couple into the "solving circle" (Glasser, 1998), which can be used as a battle-buddy type of system. This system bears the motto: "It's you and me against the problem." Thus, the soldier who incorporates the core values of the military usually finds marriage counseling with choice theory and reality therapy familiar, inclusive, and welcoming

CASE STUDY

Frank and Sara entered into counseling after 10 years of marriage. Frank is a career soldier in the military and Sara works full time and attends school full time. They have a 6-year-old child. Frank deployed twice to Iraq during their marriage. During both of Frank's deployments, Sara took over his previous household responsibilities. His former jobs at various military posts consisted of long hours training recruits. However, his current job allows him to leave work early and be off on weekends. He knows this job is temporary and he will be working long hours again in the near future. His daily routine consists of performing his military duties, picking up their son from school, driving home, and playing a computer game until his wife gets home from work. Sometimes he plays the computer game until late into the night, even after Sara has gone to bed. Sara drives their son to school, goes to work, and attends classes in the daytime. After work and school, she comes home, cooks, does homework, and goes to bed.

Sara likes to cook but the enjoyment of cooking has deteriorated because the demands of running a household; working full time, and going to school have taken a toll on her. Sara feels like she is doing everything to run the household while her husband spends his free time playing a computer game.

Frank indicates that the main reason he called for counseling is that Sara threatened him to get counseling "before our anniversary or the marriage is over." Frank claims that she threatens leaving him often and that "she says the same things over and over." The repetitive threats from her are frustrating to him because, he says, "I got it!" He feels that he does make changes in the relationship but that she says it is never enough. He indicates that at times, "it is awesome between us, but lately it has been 50% good and 50% bad."

Frank also revealed that he recently lied to his wife about looking at pornography on the Internet. In a counseling session, he stated, "I have never lied before to my wife and I pride myself on being honest." Because he wanted to be honest in the relationship, he revealed his indiscretion. He stated that he lied to her because he knew how she would react if she knew that he was looking at pornography. He has been sleeping on the couch for 2 weeks because his wife has not forgiven him and she does not trust him now. When asked if he might be suffering from PTSD, Frank said, "Some of the stuff you do over there affects you." He believes that their relationship has gotten worse since he returned from Iraq.

Sara experiences a lot of pressure from all the demands of work, school, and being a wife and mother. She feels that all Frank does at home "is play games on the computer" even though they had previously agreed on sharing household chores, including cooking. Sara says that she does not mind keeping the house clean and doing all the duties required to run the household, but feels that her husband should contribute more. Therefore, she "reminds him often" that he is part of the family. She considers him a good father when he does play with their son. Sara indicates that their social life and sex life are "lacking" and that she feels alone in her marriage.

Treatment Goals

The most important beginning point is to understand what the couple wants and expects. If the couple does not want to continue in the relationship, then counseling is usually of little value. It is vital for the couple to be committed to the relationship. Equally important is for the couple to understand that the counselor cannot solve their problems; a counselor can only educate and offer tools to the couple so they can resolve their problems. The counselor's role is to facilitate the couple's solution of their problems. This can only occur if the couple is committed to the relationship and wants to make changes to improve their relationship. In Frank and Sara's case, it was important for them to realize that they were responsible for their own happiness. It was also important to understand the expectations they had of each other, as well as to understand the perceptions and the expectations of the other person.

Assessment

Frank and Sara are trying to control each other by using what Glasser (1998) calls the "seven deadly habits": punishing, complaining, blaming, threatening, nagging, criticizing, and bribing. Frustrated and angry, both Frank and Sara enter into counseling pointing fingers and blaming each other for their unhappiness. Even so, when asked if they want to stay in the marriage, both emphatically say "yes." When asked if they want to learn ways to help their marriage, they both ardently agree again. Often unique to military relationships, the counselor must consider the signs and symptoms of combat operational stress problems that might be impeding relationship interactions.

The first three sessions comprised assessing the marriage. Within assessment are the teaching components that explain the difference between choice theory and external control (Glasser, 1998). Also, an introduction of the five basic needs, the solving circle, and the caring and deadly habits is discussed (Glasser & Glasser, 2000). The therapist or counselor explores the desire of the couple to invest in their marriage and their level of commitment for change (Wubbolding, 1988, 2011) and introduces the couple to the idea of the marriage as a separate entity (Glasser, 2005). This is similar to having a battle buddy in the Army participating in an exclusive mission. In addition, the counselor educates the soldier on combat or operational stress sources and their mechanisms or causes of stress injury. This helps to identify where symptoms are located on a stress continuum model. If these are identified, treatment may help the soldier reintegrate into the relationship (U.S. Army Public Health Command, 2010).

In reflecting on their initial courtship, Sara revealed that Frank "walked miles just to visit me at work and this impressed me very much. No one had ever worked so hard to date me and I loved the attention!" Frank says that he enjoyed their weekend dating trips. "We would just get in the car and drive. We spent the night wherever we wanted and it made me feel adventurous and free." When confronted with the last time they had fun together, both Sara and Frank could not remember when they had gone out and had a good time together. The counselor asked if they might consider going on a date the following week. When asked about how they were interacting with each other, they both admitted to saying snide comments and jumping to conclusions.

The counselor introduced the caring and deadly habits (Glasser & Glasser, 2007) and asked that they both give up one deadly habit (of their own choice) and replace it with a caring habit for a week and see if it made any difference in the way they interacted with each other.

The deadly habits are behaviors that have a basis in external control such as criticizing, nagging, and threatening, whereas the caring habits revolve around choice theory and involve behaviors such as listening, supporting, and trusting. Frank also confessed, "I'm lazy; I don't really think about cleaning up or doing anything extra because Sara does it all. I can just hide in my room and play on the computer, so I do." The couple was given the basic needs strength profile assessment (Glasser, 2005) to fill out as a homework assignment. The basic needs strength profile (Appendix 6.1) is based on Glasser's five basic needs and allows clients to self-evaluate their motivations that define personal happiness. In addition, the counselor assigned a fair fighting handout with 10 rules on it (Appendix 6.2). Instructions were to review the rules, cross out the ones they did not like, and create one that had meaning and value for them as a couple. Each partner was asked if he or she was willing to commit to focusing on one of the caring habits for the following week in order to help the marriage.

When the couple returned with the basic needs strength profile assessment, the counselor asked, "Did you learn anything from the assignment and did finding out your partner's basic need strength profile surprise you?" This allowed more discussion about the basic needs and introduced the next lesson of effective and ineffective ways of meeting basic needs. The counselor can use the basic needs strength profile with other learning tools to enhance and strengthen the meaning of basic needs.

Deployment of a soldier and separation of a family can result in feelings of anxiety, anger, depression, and even physical illness in the spouse that is left behind. The soldier that is deployed or separated may also experience feelings of anxiousness, depression, and guilt. Tanielian and Jaycox (2008) propose that the changes in skill and achievement for the left-behind spouse may be an obstacle in marital satisfaction upon the soldier's return.

Barry and Jenkins (2007) served as care providers while deployed in Afghanistan. They found a drastic need to offer relationship counseling to deployed soldiers in a group format that promoted personal responsibility and relationship-building techniques as two of the ways to enhance personal relationships. Aware of these specific relationship needs, the counselor working with a military couple can explore the different perceptions of marital roles. This can include how the couple has adapted to the predeployment stress, deployment, and reentry after the soldier returns home.

Using the basic needs strength profile, the soldier is asked to describe the basic need strength profile difference, if one exists, during deployment or separation as opposed to current strength when they are together. Likewise, the spouse is given the same task. The differences of the family unit are evaluated from when the soldier was deployed to after the soldier retuned. Did certain changes in the basic needs strength identifiers arise through necessity? Did certain major jobs of running the household change? Were some or all of the jobs and responsibilities surrendered upon the return of the deployed soldier?

Frank disclosed that he had performed particular chores before deployment (the family finances) that his wife absorbed when he deployed and that they had found that she enjoyed doing them and was more efficient at doing them. When he returned from deployment, they both decided that she would continue performing that chore.

The next three sessions offered creativity and role-playing as its foundation in order to unfold the concepts of the quality world and total behavior (Glasser, 1998). In addition, it helped the couple use their cumulative skills to negotiate their differences (Glasser, 1998). Discovering the value and meaning that the couple places on the information they share provides an opportunity for insight. This insight can help the couple solve problems. In these sessions, they role-played problem-solving skills, using a question format that explores the spouse's quality world and connects it to the basic needs. In the counseling session, Frank disclosed, "We didn't know how to 'fight' and now we are more aware of how to find out what the other person is thinking. This allows us to understand the other person's position and to figure out how to move forward." Sara added, "We used to let things go during the week and by the weekend we would just explode. Now we don't."

The couple entered the eighth session and sat together on the couch for the first time since the beginning of counseling. When asked about how they had been meeting their needs in the past couple of weeks, Sara said, "Frank took such good care of me last week when I was sick that I realized, deep down, how much he loves me. He gave me breakfast in bed, supper in bed, and made sure I took all my medications." Frank was just as quick to say, "We have been on dates to see movies and have plans to go to the mountains this weekend. We rented a cabin to have room for our little boy to play."

During the last few sessions, the couple revealed that Frank had "come down on orders" to relocate. The counselor explored the concepts of the challenges that relocation meant to them as a couple in relation to their basic needs. The goals of counseling of the military couple in the last three sessions are to help them:

- Reinforce their friendship (Glasser & Glasser, 2007; Gottman & Silver, 1999)
- Emphasize the importance of the caring habits (Glasser, 1998) to their portfolio of relationship-building tools
- Expand total quality world questions (Glasser, 1998; Wubbolding, 1988)
- Learn the components of implementing a plan (Wubbolding, 2000, 2011)

DISCUSSION

Choice theory and reality therapy offer the military couple concrete tools to address challenges that are often unique to the military marriage. Despite the challenges military marriages encounter, the National Healthy Marriage Resource Center (Hull, 2006) found that married soldiers function better and are promoted faster than those who are not married. Married soldiers are also more likely to stay on active duty. The military community has increased the number and range of programs of services for Army families. However, the military operational demands have also increased. Mental health services have had to be expanded to involve civilian mental health providers. The special nature of marriage counseling for service members is challenging in many ways not encountered in the civilian population. Additionally, many of these military marriages involve our brothers, sisters, mothers, daughters, teachers, doctors, and children.

According to Glasser (2005), the symptoms that bring people to counseling usually involve presently unhappy relationships, and marriage happens to be the most intimate of relationships. In their book, *Eight Lessons for a Happier Marriage,* Glasser and Glasser (2007) lay out a blueprint for applying the methods and techniques of choice theory (Glasser, 1998) toward a resilient and flourishing marriage. The concepts of personal responsibility, external versus internal locus of control, the basic needs, total behavior, total quality world, deadly versus caring habits, and the solving circle help direct the motivated couple toward a happy relationship.

The special qualities of the military community present complex challenges to counselors. The unique and interactive methodologies utilized in choice theory and reality therapy not only can be utilized to strengthen a struggling and challenged marriage but also can actually promote success for the family unit and, ultimately, the military unit in which the soldiers are stationed.

REFERENCES

Barry, M. J., & Jenkins, D. M. (2007, June). Relationship 101: Couples therapy in theater. *Military Medicine, 172*(6), iii–iv.

Defense Centers of Excellence (DCoE). (2010a). Building bridges: Focus guide 1: Marriage and relationship enhancement. Retrieved from http://www.dcoe.health.mil

Defense Centers of Excellence (DCoE). (2010b). Mission statement. Retrieved from http://www.dcoe.health.mil

Glasser, W. (1998). *Choice theory: A new psychology of personal freedom.* New York, NY: HarperCollins

Glasser, W. (2005). *Defining mental health as a public problem: A new leadership role for the helping professions.* Chatsworth, CA: William Glasser Institute.

Glasser, W., & Glasser, C. (2000). *Getting together and staying together: Solving the mystery of marriage.* New York, NY: HarperCollins.

Glasser, W., & Glasser, C. (2007). *Eight lessons for a happier marriage.* New York, NY: HarperCollins.

Gottman, J. M., & Silver, N. (1999). *The seven principles for making marriage work: A practical guide from the country's foremost relationship expert.* New York, NY: Three Rivers Press.

Hull, E. L. (2006). *Military service and marriage: A review of research.* National Healthy Marriage Resource Center. Retrieved from http://www.healthymarriageinfo.org/docs/review_mmilitarylife.pdf

Kane, T. (2006). *Who are the recruits? The demographic characteristics of U.S. military enlistment, 2003–2005.* Washington, DC: Heritage Foundation.

Mansfield, A. J., Kaufman, J. S., Marshall, S. W., Gaynes, B. N., Morrissey, J. P., & Engel, C. C. (2010). Deployment and the use of mental health services among U.S. Army wives. *New England Journal of Medicine, 362*(2), 101–109.

Martin, J. A., & McClure, P. (2000). Today's active duty military family: The evolving challenges of military family life. In J. A. Martin, L. N. Rosen, & L. R. Sparacino, (Eds.), *The military family: A practice guide for human service providers* (pp. 20–21). Westport, CT: Praeger Publishers.

Tanielian, T., & Jaycox, L. H. (Eds.). (2008). *The invisible wounds of war: Psychological and cognitive injuries, their consequences, and services to assist recovery.* Santa Monica, CA: Center for Military Health Policy Research, the Rand Corporation.

U.S. Army Medical Command. (2009). *Mental health advisory team (MHAT) VI Operation Iraqi Freedom 07-09.*Office of the Surgeon Multi-National Corps Iraq and Office of the Surgeon General. Retrieved from http://www.armymedicine.army.mil/reports/mhat/mhat_vi/MHAT_VI-OIF_Redacted.pdf

U.S. Army Public Health Command. (2010). Guide to coping with deployment and combat stress. Retrieved from http://chppm-www.apgea.army.mil/

Wubbolding, R. E. (1988). *Using reality therapy.* New York, NY: Harper & Row.

Wubbolding, R. E. (2000). *Reality therapy for the 21st century.* Philadelphia, PA: Brunner-Routledge.

Wubbolding, R. E. (2011). *Reality therapy: Theories of psychotherapy series.* Washington, DC: American Psychological Association.

APPENDIX 6.1: BASIC NEEDS STRENGTH PROFILE

Survival	Love and Belonging	Power	Fun	Freedom
Conservative	Social	Sets goals	Celebrate life	Independent
Cautious	Helpful	High standards	Laugh frequently	Self-reliant
Predictable	Compassionate	Achievement oriented	Sense of humor	Boundless
Focus on safety measures	Caring	Determined	Witty	Never bored
Prepared for emergencies	Kind	Driven for success	Playful	Risk taker
Responsible	Deep feelings	Challenges self	Enthusiastic	Creates options
Loyal	Hospitable	Enjoys competition	Amusing	Welcomes change
Daily exercise	Peacemaker	Values recognition	Light-hearted	Limitless
Stable, long-term relationships	Family focus	Strength of conviction	Enjoys art/ music	Adventurous
Not a risk taker	Joiner	Constant improvement	Enjoys parties/ festivals	Refuses restraints
Long-term planner	Cooperative	Committed	Joyful	Liberated
Values tradition	Diplomatic	Triumphant	Spontaneous	Unrestricted
Concern for financial security	Patient	Persistent	Cheerful	Autonomous
Adequate shelter secure	Trustworthy	Purposeful	Positive outlook	Inexhaustible

Continued

Survival	Love and Belonging	Power	Fun	Freedom
Disaster plan in place	Healthy intimacy	Competent	Seeks variety	Seeks novelty
Invest wisely	Mediator	Courageous	Seeks novelty	Own best friend
Saver	Generous	Confident	Spirited	Flexible
Organized	Friendly	Winner	Engaging	Creative
Punctual	Affectionate	Indefatigable	Game- player	Relaxes easily
Dependable	Empathic	Seeks attention/ fame	Innovative	Peaceful
Knows self-defense	Outgoing	Decisive	Active in sports	Immune to rules
Regular physical appointments	Collaborative	Seeks proficiency	Enjoys comedy	Seeks alternatives
Keeps regular routine	Team player	Infallible	Never bored	Not time- restricted
Gets adequate rest	Nurturer	Controls destiny	Love for learning	Detached
Family planning		High self-esteem	Highly creative	Unobstructed

Directions: Circle the words that best describe you. Total the number in each list.

APPENDIX 6.2: FAIR FIGHTING RULES

1. We will not fight in front of the children or in public.
2. We will give each other our full attention.
3. We will give each other a "stop" signal (agreed upon beforehand) when the fight becomes too heated and respect the signal.
4. We will not start a loaded discussion knowing the other person has to go somewhere.
5. We will not bring up past situations. We remain in the present.
6. We will give each other a warning that our intention is to ventilate feelings.
7. We will express irritations directly (using "I" statements) rather than being devious and indirect.

8. We will compare pictures of the problem to see if there are any similarities and listen for differences.
9. We will agree to disagree and respect that each other's opinion may be different from our own opinion.
10. Closure of a fight is done with a handshake, respecting that two people ended a discussion. No one person is a winner.

The relationship is more important than the fight.

7

RESOLVING DIFFERENCES, DISAGREEMENTS, AND DISCORD

Nancy S. Buck

INTRODUCTION

Ah, the joy of being in a coupled relationship! When it is good, your life seems fuller, better, and more meaningful. Even the small moments shared add to your daily joys: smiling together over a shared joke, hearing a remembered song, reigniting a passion, or simply seeing a full moon when you are separated and knowing your partner is also naming that the *hot tub moon*. Day by day, activity by activity, memory by memory, you and your partner are developing shared quality world pictures together. This also means you are sharing and building mutually need-fulfilling experiences. Happily, experiences build upon each other, adding even greater depth to your relationship.

But, oh my, the challenge of being in a coupled relationship! What happens when you and your partner have two different ideas about a similar goal? If you want to paint your bedroom a pale blue and your partner is set on a bold gold, you may compromise by painting the walls beige and using blue and yellow accents. Or perhaps you will both agree to a neutral green, a color you both like.

What if your differences involve more crucial and significant areas in your lives? Are you a spender and your partner a saver? How do you work out that difference? Are you Jewish and your partner Christian?

Perhaps you work out these differences by agreeing to observe both your own and your partner's faith respectfully. This may ultimately add to the depth of each person's faith and relationship simply by learning about the partner's religious and spiritual point of view.

The challenge of any relationship is not how to develop, grow, and mature when you agree and see eye to eye about life. The challenge is what happens when you disagree. There are bound to be more than a few moments in any relationship when each person disagrees and differs with what the partner wants. During these moments, one person, wanting what she wants, typically and frequently may use external control over her partner, trying to get him to want what she wants him to want. At the same time, and in turn, he may use external control over her to try to get her to want what he wants. There are many creative and destructive ways that one partner can attempt to control the other partner but, ultimately, all of these deadly habits diminish and may ultimately destroy loving relationships (Glasser, 2000).

THE BEGINNING OF A BAD HABIT

Let's look at a dialogue in a hypothetical case between saver wife and spender husband:

She: Honey, I've made you a meatloaf sandwich to take for your lunch today. [She's thinking that he does not need to spend $8.00 every day for a deli sandwich at work.]

He: Wonderful, thanks! I'll make reservations at the French restaurant for Friday night's dinner. With all the money we're both saving by bringing our lunch to work every day, we shouldn't have any trouble paying the restaurant bill. [He's delighted they can reward their bag lunch frugality by enjoying a lovely and expensive meal out at the end of their work week.]

Although, at first blush, this may sound like a relatively innocuous dialogue, many marriage counselors can tell stories where this first kind of dialogue, continued over time, turns into a much louder, more hostile, angrier, and meaner one. Even at this early stage of difference and discord, this couple could benefit from coaching or counseling, where they could learn to take more time to fully discuss what she means and wants from saving and what he means and wants from spending. Together, they could create a shared picture that includes both saving and spending instead of the subtle external control they are each attempting to exert over the other. Together, they could make a plan for how they would handle the financial challenges in their shared life, including a plan for surprises and unexpected events. Together, studying choice theory or undertaking couples counseling or coaching can help them develop a plan and process for working out their differences and disagreements.

Now imagine that this couple begins a family. This may be the time when differences develop into loud disagreements and discords as each works to control the other, trying to get what he or she wants. In fact, couples who are relatively happy and able to cope easily with their differences before the arrival of children may be particularly unprepared when faced with parenting disagreements and differences. They may not have known that their differences even existed until they reach the parenting point in their relationship (Buck, 2000).

CASE STUDY

To illustrate, let me give an example of one couple's parenting differences that began even before they became pregnant. Mary, the wife, was a fully conscious, fully engaged, fully educated mother-to-be, with very clear ideas and quality world pictures prepregnancy, during pregnancy, through home delivery with a midwife, and postpartum. These ideas included her life, her body, and her health, and her husband and his body and health, as well as his taking full parental work leave. Happily, the husband, John, was on board and willing to be educated to the whole process by his wife, which enabled this couple to develop a unified quality world picture of the many aspects of becoming first-time parents.

But during the second month of the pregnancy, their discussion returned to delivery. Now that the delivery was becoming much more real for John, he had different ideas and concerns about a safe and healthy delivery for his wife and child than what he had agreed to originally. John wanted his wife under the care of a medical doctor with plans for a hospital delivery to ensure all precautions and safety measures available for any unforeseen complications. The health and welfare of his wife and child were of paramount importance to him.

From Mary's perspective, John had suddenly changed what they had previously agreed upon. They had talked about using a midwife and a home delivery and had even chosen which city hospital they would choose if going to the hospital became a necessity. Using a medical doctor and hospital was only to be done in the direst of circumstances. Now it felt like John was backing out of their agreement. And with Mary's information, she felt as though John's decisions were putting both the baby and herself at greater risk. The reason she had worked so hard with

John before the pregnancy was to avoid this very conflict. And yet she found herself pregnant, with a partner whose ideas about delivery were different from hers.

Feeling hurt and angry, Mary asked for my help. This couple had not gone too far down the unhappy path of using many of the externally controlling behaviors to try to change the other person's mind. Our work together was brief and effective, and it became a process they would be able to use as they continued to face many more differences when parenting together.

Assessment

Mary and John were both anxious to come to counseling. Each was aware that he or she held very different positions about the delivery of their child. Each was aware that this was not a time in their relationship where they could agree to disagree. Delivery was going to happen. They both wanted to be a united team, agreeing on the where and how of the delivery. Each expressed concern about this being only the beginning of their parenting lives together. If they could not handle this challenge as a mutually satisfying experience, were they not in for a pretty rocky and conflicted road ahead?

I was impressed and excited to work with this couple. Too often, I am faced with couples and families who are much further down a destructive and disconnected path. This couple wanted to avoid what they described as "getting in too deep" with disharmony and discord between them. John explained that they had both experienced the ugly pain and hurt as they each watched their parents' marriages fall apart. Early in their marriage, they had agreed to seek help before it was too late, with too much hurt and anger between them.

It was also apparent that they were each concerned with issues of safety and security, the psychological aspects of the biological need for survival (Buck, 2000). A power issue involved some minor conflict around whose perception of *safe and healthy* would prevail. Luckily, their strong desire to stay connected to one another, their psychological urge for love and belonging, was driving them to work out their differences. Neither was interested in attempting to win at the expense of the other and of their relationship.

Treatment Goals

Although the obvious treatment goal for this couple was to develop an agreed upon plan for delivery, I also offered another possible goal. Together, they could learn and develop a process to work together when they faced disagreements and differences. This process would become their map and guide for any future disagreements when agreeing to disagree was not a viable solution. Both enthusiastically agreed to add this goal.

The Process

The first part of our session was to solve the immediate problem of reaching an agreed plan for delivery. With this in mind, I asked each what she and he wanted—the quality world picture for each.

From a reality therapy perspective, rather than focus on Mary's hurt and angry feelings about John changing his mind about the delivery or on John's insistence that the delivery occur within a hospital with an obstetrician present, I focused on what each person wanted. Because all behavior is purposeful and the purpose of all behavior is the person's best attempt to get what he or she wants to meet a need satisfactorily (Glasser, 1998), I knew that both Mary's hurt and John's insistence were simply behaviors that each was using to get what she and he wanted to meet a need.

Articulating and Understanding Wants   I also helped facilitate their communication to and understanding of each other. When Mary explained she wanted a safe, healthy, fully conscious home birth with a midwife and her husband present, I encouraged John to ask questions, probing for specifics so that he truly understood what Mary meant. He needed to understand her picture so clearly that he could explain it and even advocate for it if necessary.

Then it was Mary's turn to listen and understand, probing for clarity and specifics when John explained that he wanted all modern medical and technical resources ready and immediately available in case either his wife or child needed medical assistance at any point during the delivery process. While John talked, Mary's job was not to interrupt or dispute John's ideas or position with different information or statistics that made John's quality world picture wrong. Instead, she needed to understand John's position, picture, and ideas so well that she could explain them and even advocate for them if necessary.

Once they had successfully completed this first and critical step of thoroughly understanding what each wanted, the next step was to find areas of agreement within their individual pictures (Morris & Morris, 2003). Very quickly, they each realized that both wanted a safe and healthy delivery for mother and child that would provide a greater sense of safety for John as well. Both also realized that their desire for protection of their child and the other was another shared part of their picture. What became obvious then was their differing definitions of safe care for delivery.

Self-Evaluation   Before moving forward, I asked Mary and John if there was anything different about the previous process from what occurred in their normal and usual discussion when trying to solve a problem. Without hesitation, they each admitted that during their usual process, each was in such a hurry to explain his position, her idea, and why her idea or his position was the correct one and why the other person was wrong that they rarely fully appreciated and understood the spouse's perspective. Each was interested in being right and proving the other wrong. To really listen for understanding, working to become an advocate for the other person's position was very different, eye-opening, and almost exciting.

What Are You Doing? What Will You Do?   Next, the questioning focused on asking them how they wanted to proceed in developing a unified definition for safe care for delivery. From a choice theory perspective, I was asking each what his or her perception was of the existing, newly defined problem (Morris & Morris, 2003). John admitted that Mary had pursued and researched different and additional information about delivery, including home delivery, birthing centers, and hospital birth. He had listened to all she explained to him, but he was using his own male instincts and, perhaps, fears. But because we lived in modern times, he figured, should he not advocate for using modern advantages that ensured the greatest safety? However, if he really was going to attempt to hold that his position was the better position, he needed to do more research, which he promised he would do. And once he had more solid and factual information, he would bring that to Mary to continue their discussion.

Pleased with John's willingness to investigate and research both sides of the issue, Mary agreed that she would investigate, for a compromise position, some delivery alternative that was between total home birth and hospital birth. This meant she would investigate and research more information about birthing centers. She would also line up the names of several midwives for them to interview, being sure that each was associated with an obstetrician. John then agreed to begin a list of questions they would ask the potential midwives, especially including his concerns about safety, health, and modern medicine.

As is so often the case, once this couple better understood and appreciated the quality world picture of the other, they were able to use their competent problem-solving skills to work together toward finding their mutual quality world picture that included aspects of each other's individual quality world pictures. There was no need for me to ask either to self-evaluate the effectiveness of their previous behavior in attempting to solve their problem. Each had already self-evaluated and knew that what he or she was doing was not helping him or her get what was wanted and was not helping their relationship. Working together to solve the problem, each knew that the other had a sense of good will and felt concern for the health, safety, and well-being of the other. This concluded their initial session with plans to return in 2 weeks, after they had each done their homework.

Planning   As it turned out, John called and postponed their session for 2 weeks, explaining that they were in the midst of interviewing midwives. They wanted to return for their final session with their plan intact. They arrived happy and satisfied. After John did his research, he realized that Mary was correct in her assessment of a home birth using a qualified midwife being in fact statistically safer for his wife and child. They had toured birthing centers in town, but were not happy that the delivery professional that would be present for their child's birth was not guaranteed because the staff worked on a rotation schedule. The feeling of home was better at the birthing center when compared to the birthing room at the hospital, but it might not be available for their birth because the two rooms were on a first-come, first-served basis. Both agreed that their own home with a guaranteed known midwife present at birth was preferable to the birthing center. John was the most surprised and pleased at his changed mind.

They had interviewed three midwives and were happy to find that they both had similar feelings about each one. One was too medical; they felt that there was not much difference between her and an obstetrician. The second was a bit too casual, explaining that she had handled some births on her own where others might have gone to the hospital instead. This was too close to John's concerns and fears. The midwife they chose felt like a perfect match and fit. She was associated with a medical doctor on the list of doctors they wanted to use if necessary. Both Mary and John felt confident and reassured under her guidance and care.

Developing a Long-Term Problem-Solving Process   Happy and satisfied that they had accomplished their first goal and developed a delivery plan, Mary and John were ready and anxious to develop a process to work out their differences together. Together, we reviewed what they had done:

- Wants and quality world picture
  - Each person clearly and specifically explains what he wants. Avoid any of the deadly habits directed toward the other. All hurt, angry, or negative feelings are to be redirected back into explaining what the quality world picture is. Remember that all behavior is purposeful, so all of the deadly habits and all expressions of negative feeling are simply the person's best attempt to get what she wants. For the sake of the relationship and the friendship, it is better to state what you want and avoid using the deadly habits or expression of negative feelings.
  - The partner's job is to listen and seek clear understanding, asking questions for clarification if necessary, but not asking questions to correct or disagree. The partner is seeking to understand the other's quality world picture so well that she could explain it to another, even advocate for this position as if it were her own.
- Find the shared wants.
  - After each person has articulated his or her wants, together they identify the shared quality world aspects, creating a new shared quality world picture.
  - The areas of disagreement and differences in perceptions become identified.
- Seek information, seek shared perceptions.
  - Each person identifies what he or she will do to research and seek more information about the areas of disagreement and differences. Are these differences a matter of opinion or based on concrete evidence and facts?
  - Seeking more information to better understand the other's position as well as validating or discrediting one's own position is the goal.
- Make a plan or plan to wait:
  - With the additional information, can a plan be made that satisfies both people and the agreed upon shared quality world picture? If the answer is no, agree to delay making a plan for an agreed upon time. Commit to the other that while you are in limbo, neither of you will attempt to try to convince or change the other person's mind or position. Commit to avoid using the deadly habits. Instead, ask yourself or your spouse, "What do you want?" or tell your spouse, "This is what I want."

- Continue to gather more information as appropriate.
- Make a plan as soon as possible.

This marked the successful conclusion of our work together. As a footnote, John and Mary were in touch with me 9 months following the safe and healthy home delivery of their daughter, Jane. They explained that they were now in the midst of trying to come to an agreement regarding vaccinating their daughter; each held an opinion and position different from the other's. They were using their agreement process. Mary explained that she was relieved to have this process to help them both. Although it was time consuming and sometimes arduous, she felt certain that using this process was keeping them from inadvertently hurting or disrespecting the other and was ultimately making them a stronger couple and family.

DISCUSSION

The statistics of marriages that end in divorce is a testament to how difficult it is for people to learn to get along successfully. Today's families are more diverse than ever, making the challenges even greater. It is no longer extraordinary to have a family where original parents divorce, and then each marries a new spouse, creating households of children that include "yours," "mine," and "ours." With these new blended families, the possibility for differences and disagreements between any and all members of the new family are unlimited. Just because the newly married adults like and love each other is no guarantee that the new stepsiblings will love, respect, or even like each other or their new stepparents. Do couples, siblings, or families know how to disagree with one another while still respecting the needs of the other family members? Does the family have a process or practice to handle such discord successfully and amicably without using any of the deadly habits? Helping families learn, practice, and use a process of peaceful problem solving provides the family with the essential skills to handle their differences, difficulties, and discords (Buck, 2009).

When a couple understands choice theory, the desire for one partner to want to change the other does not vanish, but partners learn to act differently. Instead, differences and discords can lead to greater understanding, appreciation, and depth as the couple works together and, amicably and respectfully, reaches a resolution satisfactory to each person. Learning and developing a process of how a couple will resolve their differences is a powerful skill that can be learned. Implementing the process of reality therapy helps couples learn the skill of resolving their differences—the key that helps relationships develop into the exceptional ones in today's culture. These are the strong, mature, respectful, need-satisfying, and lasting couples who choose to stay together—not out of necessity, but rather out of friendship, desire, and choice.

REFERENCES

Buck, N. S. (2000). *Peaceful parenting*. San Diego, CA: Black Forest Press.

Buck, N. S. (2009). *Why do kids act that way? The instruction manual parents need to understand children at every age*. Charlestown, RI: Peaceful Parenting Inc.

Glasser, W. (1998). *Choice theory: A new psychology of personal freedom*. New York, NY: HarperCollins.

Glasser, W. (2000). *Getting together and staying together*. New York, NY: HarperCollins.

Morris, S., & Morris, J. (2003). *Leadership simple. Leading people to lead themselves*. Santa Barbara, CA: Imporex International Inc.

8

CAREGIVING COUPLES AND ADULT CHILDREN DIAGNOSED WITH MAJOR PSYCHIATRIC DISORDERS

Willa J. Casstevens

INTRODUCTION

Caregiving couples experience high levels of stress and, when the care recipient is involved in the mental health system, this can compound the stress involved. Parent caregivers in such situations face unique challenges in navigating a fragmented and under-resourced mental health system. This is especially relevant when the adult child, the care recipient, has been given a major psychiatric disorder diagnosis. When this occurs, society in general and mental health professionals in particular may expect parents to manage their adult child's behavior. As parents struggle to meet these expectations by attempting to control the adult child's behavior, family relationships deteriorate, sometimes to the point of estrangement. The stress involved can bring caregiving couples into counseling at times of crisis and can put them at risk of separation or divorce.

A reality therapy and choice theory approach to counseling has much to offer these couples and families. In addition to educating a couple about choice theory, the counselor using a reality therapy approach can also educate them about Glasser's (2000, 2003) perspective on psychiatry and mental health and can help couples and families come to a basic understanding of both (Wubbolding, 2000, 2011). This is increasingly important in the context of deinstitutionalization and biological psychiatry in America.

Over the last 60 years or so, American mental health policy has supported what is known as deinstitutionalization—a process that peaked in the 1970s and 1980s (Mechanic & Rochefort, 1990). Deinstitutionalization during that period involved discharging large numbers of state hospital patients to the community, and it was associated with allocating federal funds for community mental health centers (Mechanic, 2008). The locus of treatment and care for individuals labeled with major mental disorder diagnoses shifted to community-based organizations, with in-patient psychiatric hospitalization only if behaviors were a danger to self or others. Community treatment and services, however, remain fragmented and can be difficult to navigate, even with the aid of case managers (Solomon, 1998). Caregiving responsibilities fall to a large extent on parents and other family members, even if the adult child does not reside in the caregiver's household (Benson, 1994). Estimates indicate that approximately one third to two thirds of adults diagnosed with major psychiatric disorders live with family members or receive primary care from them (Solomon, 1996).

VIOLENCE AND VICTIMIZATION

A concern that has been much publicized is that of violent behavior by individuals who are diagnosed with a major mental disorder (e.g., the Virginia Tech shootings in 2007). The lifetime prevalence of violence among individuals diagnosed with a major mental disorder, including those who do not live with family members, is slightly over twice that of individuals who are not diagnosed with a major mental disorder (16% and 7%, respectively; Friedman, 2006). Although the lifetime prevalence of violence is infrequent, when it does occur it tends to receive a great deal of media attention.

Swartz et al. (1998) found that, among involuntarily hospitalized adults diagnosed with major mental disorder, those between 18 and 29 years old, and those who were male, were more likely to have committed a violent act within the previous 4 months of intake than were older adults or females. While African Americans in this sample were more likely than Whites to have behaved violently, it was determined that this was due to social environmental and socioeconomic factors rather than to race or ethnicity (Silver, 2000; Swanson, Holzer, Ganju, & Jono, 1990; Swartz et al., 1998). That young men behave violently more often than do older adults or women is not, perhaps, surprising: Glasser (2003) observed, "By far the most dangerous people in any community are unhappy young men between the ages of eighteen and thirty" (p. 18).

In examining violence, victimization, and caregiver burden, research suggests that caregiver monitoring exacerbates the association between violence and caregiver financial burden (Thompson, 2007). This fits within the framework of choice theory: When external control (monitoring behavior) increases, then relationships deteriorate and the potential for violent acting-out behavior tends to increase. If violent behavior actually occurs, this in turn would tend to increase caregiver financial burden because of costs associated with household damages, legal costs, personal injuries, and/or psychiatric hospitalization. Violence, of course, is not necessarily one-way, and "family members and relatives often are identified as the perpetrators of acts of violence against their relatives who have mental illness" (Marley & Buila, 2001, p. 122). Based on an extensive review of the literature, Solomon, Cavanaugh, and Gelles (2005) estimated the rate of family violence in caregiving situations at between 10% and 40%, which may suggest a higher rate of violent behavior than Friedman's (2006) statistics indicate. Meaningful comparison of mental health and domestic violence statistics is difficult and sometimes impossible, however, because sampling and reporting methods vary across studies.

Associated with violence is victimization, and estimates of physical and sexual abuse histories for individuals diagnosed with major mental disorders range from 40% to over 90%, with rates depending on a sample's gender distribution (Solomon et al., 2005). Estimates of physical and sexual abuse are higher in the diagnosed population than in the general population, where reported childhood victimization ranges from 14% to 32%, depending on type of abuse and gender, and the rate of reported victimization in adulthood is 36% (Briere & Elliott, 2003). While experiencing long-term victimization is significantly associated with violent behavior, experiencing short-term victimization is not (Solomon et al., 2005). This fits within choice theory: Long-term victimization can threaten an individual's basic need of survival, as well as his or her basic need of love and belonging; an individual's basic need of power can also be threatened. It should also be noted that parent and/or other family member caregivers may not be aware of present or past victimization, even in instances when it is or was long term.

A perceived threat to the basic need of power can occur, for example, in situations where an individual is financially dependent on family; within a choice theory framework, it is not surprising that individuals diagnosed with major mental disorders who were "more financially dependent on their families were more likely to threaten others or behave violently" (Solomon et al., 2005, p. 48). The following case presentation does not address family member violence or victimization directly; however, it is hoped that this brief discussion demonstrates the applicability and relevance of choice theory and reality therapy in such cases.

CAREGIVER CHALLENGES

The multiple sources of stress experienced by couples in caregiving situations are often referred to as "caregiver burden." When an adult child diagnosed with a severe and persistent mental disorder lives with parents, caregiver burden often includes personal caregiving tasks for the adult child, coping with disruptive and/or disturbing behaviors, and the financial burden involved with support and treatment provision (Thompson, George, Swartz, Burns, & Swanson, 2000). Role strain can occur if parents struggle with their caregiving role, particularly if they experience anxiety or depression (Thompson et al., 2000)—that is, in choice theory language, if they are worrying or depressing. Role conflict or overload may occur as caregivers juggle multiple roles in the workplace, in the community, and at home. Societal and professional expectations that parents should be able to get children to comply with treatment recommendations can compound these issues.

With all this involved, it is easy to lose sight of the fact that parent caregivers may still be grieving the loss of their "ideal child" and the loss of hopes and dreams in their quality worlds that the ideal child represented. This means that, in addition to the caregiver burden and role strain and social expectations and pressures, parents may be going through a grieving process (Miller, 1996). Kübler-Ross (1997; Kübler-Ross & Kessler, 2007) identified five stages of grief among and between which people move in very fluid and individual ways. Using choice theory language, these stages are known as denying, angering, bargaining, depressing, and accepting.

Individuals take time, sometimes years, to resolve grief and loss associated with bereavement. Losses that occur solely within parents' quality worlds, however, are sometimes neither recognized nor acknowledged, making it difficult for parents to accept their adult child as he or she is in the present. Assisting a couple with recognizing and acknowledging their individual losses and letting them know that it is all right to grieve can help them to put their adult child into their respective quality worlds as he or she is now, rather than as they had hoped or dreamed he or she would be. Grieving in this context is likely to include thinking, acting, feeling, and somaticizing (i.e., total behavior).

This will be a different process for each parent, just as the losses in the quality world will be unique to each individual. While this chapter focuses on caregiver concerns, it should be noted that the adult child is also grieving losses—in the real world and his or her quality world—related to many different aspects of diagnosis, medication, daily activities, relationships, and life goals.

In order to preserve client confidentiality, the following case study is a hypothetical case with elements drawn from many families and situations with which the author has worked over the years.

CASE PRESENTATION

In their early 60s, Jack and Evelyn became caregivers for their youngest daughter, then in her 30s. They had both retired and had already downsized from a house to a large condominium. Jack had taken early retirement from the university where he was a professor. Evelyn had been a school teacher and raised three children while working full time; she retired about the same time that Jack did. The two oldest children were married and had young children; they had completed university degrees and had professional jobs out of state. Jack and Evelyn visited both children's families on a regular basis twice a year.

Jack and Evelyn came into counseling at their wits' end. Their youngest daughter, Sue, had been living with them for 2 months and they described the situation as "unendurable." They could not understand how they had "failed" or how Sue could have contracted schizophrenia or schizoaffective disorder (she had been given different diagnoses by different psychiatrists). They informed the counselor that the family had no history of mental illness or abuse of any kind.

Sue had been the "odd one" of the three siblings. She was diagnosed with attention deficit disorder in elementary school and prescribed methylphenidate for it. In middle school, she was a loner who got into trouble repeatedly and in high school she played around with drugs. She graduated from high school, but dropped out of college to marry a man of whom her family disapproved. Although she denied it, her parents and siblings believed that her husband beat her, and she had a psychotic breakdown when he abandoned her after 2 years of marriage. Sue then spent several months psychiatrically hospitalized and was discharged on antipsychotic and antidepressant medications to an outpatient treatment program and an assisted living facility. Since then, she had had several involuntary psychiatric hospitalizations because of threats of suicide that she made to staff at the assisted living facility. Sue received monthly disability checks that paid for her room and board; she was on Medicaid, which covered her psychiatrist appointments, prescription medication, day treatment program, and the case management services she received from the community mental health center.

Sue began attending the day treatment program less and less frequently and started to hang out with friends at the assisted living facility who "smoked dope and used coke occasionally," although Sue denied any illegal drug use when her case manager confronted her. Sue refused to return to attending the day program regularly, saying it was "boring and a waste of time." Sue's case manager reported this to her parents because signed consent paperwork was on file and Jack and Evelyn had been actively involved with all of Sue's treatment. Jack and Evelyn consulted with one another after the call from Sue's case manager, and they decided that Sue needed to move in with them. After an initial angry outburst, Sue agreed to the move and was now living in Jack and Evelyn's guest room.

Things went well, and Jack and Evelyn left for a week to visit Sue's brother and his family. When they returned, Sue was quiet and seemed fine. Her pill count checked out when Evelyn monitored her medications. Three days after Jack and Evelyn's return, Sue awakened the household, screaming and sobbing, at about two o'clock in the morning. Sue insisted there was a man in her room who woke her and grabbed her, but no one other than the family was in the condominium. Her parents slept together in the master bedroom downstairs and were positive no one had entered—the alarm was on and the system showed no signs of tampering. Her parents told Sue this, and after several sleepless nights due to Sue's screams and subsequent insistence they search the condo, they told her to stop attention-seeking and lying. Sue locked herself in her bedroom, which had a private bathroom, and refused to open the door for medication or food.

This had happened 2 days prior to the counseling session. Jack and Evelyn came to the initial session because they did not want to call the police and mobile crisis team to enter Sue's room forcibly, as her case manager and psychiatrist suggested. If Sue did not come out soon, though, they did not see any other option. Jack and Evelyn had asked Sue's case manager to arrange the counseling appointment, in the hope that the counselor would be able to tell them what to do with their daughter, who was "clearly out of control." A neighbor who knew Sue had offered to stay at the condominium while they attended the appointment.

CASE CONCEPTUALIZATION AND TREATMENT GOALS

From the case manager and psychiatrist's medical model perspective, Sue appears to be decompensating and exhibiting an increase in psychotic symptoms (hallucinations) that may be due to sporadic use of street drugs and/or noncompliance with prescribed medications. Sue's behavior may present a danger to self because she is refusing food and has a history of suicidal ideation. Therefore, if she can be assessed by someone with the necessary credentials and authority, she is likely to meet criteria for involuntary psychiatric hospitalization. This is the reasoning behind the case manager and psychiatrist's suggestion that Jack and Evelyn call the police and mobile crisis team to come to their home. Jack and Evelyn did not want to disagree with Sue's psychiatrist, but did express the concern that if they followed through with this recommendation, they risked shattering an already strained relationship with their daughter. They hoped that having an opportunity to problem solve with a counselor who was familiar with severe and persistent mental illness and treatment might bring alternative options to light.

In reality therapy, counseling focuses on people's basic need for love and belonging, and their present relationships; Jack and Evelyn's intuitive concern about forcing Sue out of her room and into the hospital seemed valid. They had been using external control psychology, along with the seven deadly relationship habits (criticizing, blaming, complaining, nagging, threatening, punishing, and bribing or rewarding to control) in their attempts to control Sue's disruptive nighttime behavior (Glasser, 1998; Wubbolding, 2011). External control psychology had exacerbated the situation that brought the couple into counseling and each of the family members involved was experiencing distress and unhappiness. Each person's total behavior was an attempt to rebalance his or her scales to meet basic needs. By coming into counseling, Jack and Evelyn were each trying to meet basic needs of love and belonging and to maintain cherished pictures of family and parenthood in their quality worlds. Sue's total behavior appeared to be an attempt to meet basic needs of power and freedom, without completely losing relationships with her parents that were important for her quality world and for her basic need of love and belonging.

The immediate goal was to resolve the impasse at home because this was the crisis that brought the couple to counseling. Counselor objectives included:

- educating the couple about choice theory and external control psychology and helping them start using choice theory with one another and with Sue

- introducing the couple to the seven caring and seven deadly habits and helping them start using the seven caring habits (supporting, encouraging, listening, accepting, trusting, respecting, negotiating differences) at home
- exploring Dr. Glasser's perspective on psychiatry and mental health with the couple

Because Sue's parents had already included her in their quality worlds, grief and loss were not addressed in the initial case formulation, despite the question asked during the first session: "What did we do wrong raising her?" Similarly, the couple was not at risk of separation or divorce, so this did not need to be addressed in counseling.

To accomplish the immediate goal, the counselor presented the concepts of external control psychology and choice theory to Jack and Evelyn, which allowed them to consider the situation in a new light. The counselor then introduced the seven caring and seven deadly habits and their fit with choice theory and basic needs. Jack and Evelyn already knew that what they were doing was not working. They also recognized that their use of the seven deadly habits had escalated the situation, although they had never heard of choice theory or the seven caring and seven deadly habits prior to the session. Because both enjoyed reading, the counselor suggested they might consider starting *Choice Theory: A New Psychology of Personal Freedom* (Glasser, 1998) as "homework" before their next session.

After the counselor introduced choice theory and asked the question, "Whose behavior can you control, really?" Evelyn said, "Do you think she might come out if I apologized for calling her a liar?" Jack nodded and remarked,

> You know, we never asked her if it might have been a nightmare. Remember the time she woke up crying because she actually saw a wolf in her room after that bedtime story I read her? It wasn't until I told her to tell it to go away that she could get back to sleep.

They were on their way. Sue did come out of her room after her mother apologized and, at the counselor's suggestion, was invited to join the following week's counseling session.

Evelyn and Jack bought the book *Choice Theory: A New Psychology of Personal Freedom* and began reading it aloud to one another in the evenings—something they said they used to enjoy doing with poetry. They also sat down over coffee and asked Sue if she would mind telling them about her "dream—or whatever it was." Sue was able to share that she had been scared because it reminded her of her ex-husband, who used to come home drunk and throttle her if he found her sleeping. She had never felt able to ask for help when she was married because "everyone had told me not to marry him and would say, 'I told you so.'" Other conversations happened gradually, in bits and pieces, and healed large parts of Jack and Evelyn's relationships with their youngest daughter as trust began to be rebuilt. The third session focused on helping Jack and Evelyn develop ways to use choice theory and the seven caring habits in specific situations at home so that they could avoid a return to external control psychology.

Jack and Evelyn did not want long-term counseling, and at their fourth session shared that they thought they were doing fine as a family and that this would be their final session. Evelyn had looked up more of Dr. Glasser's work online; she started reading *Warning: Psychiatry Can Be Hazardous to Your Mental Health* and talked to Sue's psychiatrist about it prior to the fourth counseling session. Evelyn asked to schedule a fifth session, which she attended alone and during which she shared that while she appreciated choice theory, she had grave doubts about a "purely psychological" view of mental disorders. She did not want to discuss this and stated that she did not want Sue to read *Warning: Psychiatry Can Be Hazardous to Your Mental Health*. This led to a brief revisiting of choice theory and the question: "Whose behavior can we control?" As it turned out, Sue had joined and was enjoying several women's groups at the condominium's community center and had told her mother that she did not need or want to read another book.

DISCUSSION

Ideally, Jack, Evelyn, and Sue would all have read *Warning: Psychiatry Can Be Hazardous to Your Mental Health* (Glasser, 2003) and discussed the various concerns and questions that came up in a family session with the counselor. Instead, only one family member read the book, and this occurred before the counselor provided any background on its content. In working with a motivated couple, both of whom enjoyed reading, this was an error on the counselor's part. Differences between Dr. Glasser's views on psychiatry and mental health and biological psychiatry's medical model should have been mentioned briefly in the second session, time permitting, and certainly by the third session.

Jack and Evelyn did, however, accept choice theory and begin using choice theory and the seven caring habits with one another and with Sue. The family was fortunate in that the basic need of survival was met in full for all family members. This financial security meant that the parent caregivers could afford to pay out of pocket for couple counseling (and the books) that the adult child's insurance would not have covered. They also had adjusted to their daughter receiving a major mental disorder diagnosis and were already coordinating caregiving treatment and services with mental health professionals in the community. This simplified counseling with the couple and family because basic survival needs, treatment and service coordination, and grief issues did not need to be addressed.

Glasser's (2000, 2003, 2005) approach to mental disorders and psychiatric treatment is quite different from that of the medical model of mental illness that became popular in the late twentieth century, when mainstream American psychiatry embraced the view that mental disorders were biologically based brain diseases. This resurgence of biological psychiatry in the 1950s had followed the development of tricyclic antidepressants and the major tranquilizer chlorpromazine (Whitaker, 2002); "major tranquilizers" are now called antipsychotics. The medical model emphasizes somatic treatment ("soma" means body), which includes psychotropic medications that aim to help a patient manage symptoms and maintain a socially acceptable level of functioning. Glasser (2000, 2003), however, does not view severe and persistent mental disorders as biological brain diseases, but instead as psychological problems—and, as one of the axioms of choice theory states, all long-lasting psychological problems are relationship problems. Reality therapy and choice theory can be used, of course, whether or not the reader agrees with Dr. Glasser's view of severe mental disorders. Similarly, reality therapy can be used in conjunction with psychotropic medication, as the case presented indicates.

Severe and persistent mental disorders include diagnoses such as schizophrenia, bipolar disorder (formerly called manic depression), and recurrent major depressive disorder. Anxiety disorder diagnoses (e.g., posttraumatic stress disorder) are sometimes also included in this category. Glasser (2000, 2003) views psychiatric symptoms as expressions of unhappiness and suggests that mental health can be seen on a continuum that ranges from mentally ill through unhappiness to mentally healthy. He notes that some people's diagnoses do meet biological criteria for brain pathology, and these include those with Alzheimer's and epilepsy, among others. Glasser (2003, 2005) argues that the severe and persistent mental disorder diagnoses, however, are not distinguished by any brain pathology or by any imbalance in brain chemistry, despite what biological psychiatrists may say. Individuals who have been given these diagnoses are often distinguished, in this author's professional experience, by profound unhappiness with life in general and with their interpersonal relationships in particular. Their basic need for love and belonging is frequently unmet, and many other basic needs may be unmet as well.

The medical model of mental illness still pervades much of American mental health, and the advocacy group, National Alliance on Mental Illness (NAMI, founded in 1979), has embraced the medical model. Because the pharmaceutical industry provides over two thirds of NAMI's funding, NAMI has been criticized for promoting psychotropic medications sold by the industry (Harris, 2009). Further drug developments, such as the atypical antipsychotics (e.g., olanzapine) and the serotonin selective reuptake inhibitors (SSRIs; e.g., fluoxetine), have contributed to expanding the pharmaceutical industry's increasingly lucrative mental health market (Angell, 2004).

Biologically oriented mental health practitioners, including psychiatrists, social workers, and counselors, often tell families and patients that psychotropic medication helps correct a chemical imbalance in a patient's brain. The difficulty with this is that the existence of such an imbalance has never been established (Boyle, 2002). The potential for harmful effects from these medications, however, is well established (Breggin, 1990; Breggin & Cohen, 2007; Gomorry, Wong, Cohen, & Lacasse, 2011; Jacobs, 1995). Regarding taking psychiatric drugs, note that Glasser (2003) clearly states,

> I do not advise you to stop it if you or your family is convinced this drug is helping you. But if you do stop it, do it slowly, because an abrupt withdrawl [sic] from these strong, brain-altering drugs may also be harmful to your mental health. (p. 19)

Because these drugs are now commonly prescribed in treatment settings, it behooves practicing clinicians to educate themselves on psychotropic medication usage and the professional dilemmas that may arise when working with clients for whom these medications are prescribed (Bentley & Walsh, 1998, 2006).

A biologically based model of mental disorders does have validity for some of the conditions in the *Diagnostic and Statistical Manual of Mental Disorders* (*DSM;* American Psychiatric Association, 2000). For example, Alzheimer's disease has been shown to involve progressive plaque and tangle buildup in the brain and a loss of connections between neurons. For severe and persistent mental disorder diagnoses such as schizophrenia, however, experts in the field continue to debate the validity of *DSM* diagnostic categories and somatic treatments (e.g., Andreason, 1984; Boyle, 2002; Breggin, 1994; Torrey, 2006). Kirk and Kutchins (2008) offer a noteworthy critique of the scientific basis claimed for the *DSM* itself (see also Kutchins & Kirk, 1997).

It is important that counselors who use reality therapy and choice theory with couples and families involved in the mental health system be aware of the information (and misinformation) to which their clients may have been exposed because many will have incorporated at least some part of this into their quality worlds. Reality therapy and choice theory have much to offer parents and other couples who are in caregiving situations with an adult who has been given a major mental diagnosis. The approach also has much to offer the care recipients of these relationships and can be used successfully with individuals or families as well as couples.

REFERENCES

American Psychiatric Association (2000). *Diagnostic and statistical manual of mental disorders* (4th ed., Text revision). Washington, DC: APA.

Andreasen, N. C. (1984). *The broken brain: The biological revolution in psychiatry.* New York, NY: Harper and Row.

Angell, M. (2004). *The truth about the drug companies: How they deceive us and what to do about it.* New York, NY: Random House.

Benson, P. R. (1994). Deinstitutionalization and family caretaking of the seriously mentally ill: The policy context. *International Journal of Law and Psychiatry, 17*(2), 119–138.

Bentley, K. J., & Walsh, J. (1998). Advances in psychopharmacology and psychosocial aspects of medication management: A review for social workers. In J. B. W. Williams & K. Ell (Eds.), *Advances in mental health research: Implications for practice* (pp. 309–342).Washington, DC: NASW Press.

Bentley, K. J., & Walsh, J. (2006). *The social worker and psychotropic medication: Toward effective collaboration with mental health clients, families, and providers* (3rd ed.). Belmont, CA: Brooks/Cole.

Boyle, M. (2002). *Schizophrenia: A scientific delusion?* New York, NY: Routledge.

Breggin, P. R. (1990). Brain damage, dementia and persistent cognitive dysfunction associated with neuroleptic drugs: Evidence, etiology, implications. *Journal of Mind and Behavior, 1,* 425–464.

Breggin, P. R. (1994). *Toxic psychiatry: How therapy, empathy and love must replace the drugs, electroshock and biochemical theories of the "new psychiatry."* New York, NY: St. Martin's Press.

Breggin, P. R., & Cohen, D. (2007). *Your drug may be your problem: How and why to stop taking psychiatric medications.* Cambridge, MA: Da Capo.

Briere, J., & Elliott, D. M. (2003). Prevalence and psychological sequelae of self-reported childhood physical and sexual abuse in a general population sample of men and women. *Child Abuse & Neglect, 27,* 1205–1222.

Friedman, R. A. (2006). Violence and mental illness—How strong is the link? *New England Journal of Medicine, 355*(2), 2064–2066.

Glasser, W. (1998). *Choice theory: A new psychology of personal freedom.* New York, NY: HarperCollins.

Glasser, W. (2000). *Counseling with choice theory: The new reality therapy.* New York, NY: HarperCollins.

Glasser, W. (2003). *Warning: Psychiatry can be hazardous to your mental health.* New York, NY: HarperCollins.

Glasser, W. (2005). *Defining mental health as a public health issue: A new leadership role for the helping and teaching professions.* Chatsworth, CA: William Glasser Inc.

Gomorry, T., Wong, S. E., Cohen, D., & Lacasse, J. (2011). Clinical social work and the biomedical industrial complex. *Journal of Sociology and Social Welfare, 28*(4), 135–165.

Harris, G. (October 21, 2009). Drug makers are advocacy group's biggest donors. *New York Times.*

Jacobs, D. H. (1995). Psychiatric drugging: Forty years of pseudo-science, self-interest and indifference to harm. *Journal of Mind and Behavior, 16,* 421–470.

Kirk, S. A., & Kutchins, H. (2008). *The selling of the DSM—The rhetoric of science in psychiatry.* New Brunswick, NJ: Aldine.

Kübler-Ross, E. (1997). *On death and dying.* New York, NY: Scribner.

Kübler-Ross, E., & Kessler, D. (2007). *On grief and grieving: Finding the meaning of grief through the five stages of loss.* New York, NY: Scribner.

Kutchins, H., & Kirk, S. A. (1997). *Making us crazy: DSM: The psychiatric bible and the diagnosis of mental disorders.* New York, NY: Free Press.

Marley, J. A., & Buila, S. (2001). Crimes against people with mental illness: Types, perpetrators, and influencing factors. *Social Work, 46*(2), 115–124.

Mechanic, D. (2008). *Mental health and social policy: Beyond managed care* (5th ed.). New York, NY: Pearson Education.

Mechanic, D., & Rochefort, D. A. (1990). Deinstitutionalization: An appraisal of reform. *Annual Review of Sociology, 16,* 301–327.

Miller, F. E. (1996). Grief therapy for relatives of persons with serious mental illness. *Psychiatric Services, 47*(6), 633–637.

Silver, E. (2000). Race, neighborhood, disadvantage, and violence among persons with mental disorders: The importance of contextual measurement. *Law and Human Behavior, 24*(4), 449–456.

Solomon, P. (1996). Moving from psychoeducation to family education for families of adults with serious mental illness. *Psychiatric Services 47*(12), 1364–1370.

Solomon, P. (1998). The conceptual and empirical base of case management for adults with severe mental illness. In J. B. W. Williams & K. Ell (Eds.), *Advances in mental health research: Implications for practice* (pp. 482–497). Washington, DC: NASW Press.

Solomon, P. L., Cavanaugh, M. M., & Gelles, R. J. (2005). Family violence among adults with severe mental illness: A neglected area of research. *Trauma, Violence, & Abuse, 6*(1), 40–54.

Swanson, J. W., Holzer, C. E., Ganju, V. K., & Jono, R. T. (1990). Violence and psychiatric disorder in the community: Evidence from the epidemiologic catchment area surveys. *Hospital and Community Psychiatry, 41*(7), 761–770.

Swartz, M. S., Swanson, J. W., Hiday, V. A., Borum, R., Wagner, H. R., & Burns, B. J. (1998). Violence and severe mental illness: The effects of substance abuse and nonadherence to medication. *American Journal of Psychiatry, 155*(2), 226–231.

Thompson, M. S. (2007). Violence and the costs of caring for a family member with severe mental illness. *Journal of Health and Social Behavior, 48,* 318–333.

Thompson, M. S., George, L. K., Swartz, M., Burns, B. J., & Swanson, J. W. (2000). Race, role responsibility, and relationship: Understanding the experience of caring for the severely mentally ill. *Research in Community Mental Health, 11,* 157–185.

Torrey, E. F. (2006). *Surviving schizophrenia: A manual for families, patients and providers* (5th ed.). New York, NY: HarperCollins.

Whitaker, R. (2002). *Mad in America: Bad science, bad medicine and the enduring mistreatment of the mentally ill.* Cambridge, MA: Perseus Publishing.

Wubbolding, R. E. (2000). *Reality therapy for the 21st century.* New York, NY: Routledge.

Wubbolding, R. E. (2011). *Reality therapy.* Washington, DC: American Psychological Association.

9

WHEN CHILDHOOD TRAUMA HAUNTS
THE COUPLE RELATIONSHIP

Gloria Smith Cissé, Terri Earl-Kulkosky, and Jeri L. Crowell

INTRODUCTION

Violence in the family strikes at the very core of what is needed to build a solid, lasting relationship. Family violence can have a long-lasting effect, not only on the persons involved in the direct violence but for generations to come. According to the National Coalition Against Domestic Violence (NCADV) (2007), children who have witnessed domestic violence have to contend with the impact of the violence, and the trauma they experience can show up in emotional behavioral, social, and physical disturbances that affect their development and can continue into adulthood. Witnessing physical attacks by parents or caregivers is a strong risk factor for transferring violence between generations. Boys, for instance, who witness this type of violence are more likely as adults to abuse their partners (Godbout, Dutton, Lussier, & Sabourin, 2009; NCADV, 2007), and 30% to 60% of those who abuse their partners also abuse their children (Edleson, 1999).

The adults involved in the violence may assume the children are not aware of what is going on in the other room, but children are all too often listening, hiding, and sometimes thinking of ways to help the abused or abusive parent. They are taking mental snapshots and placing these representations of behaviors into their mental picture albums, or quality worlds (Glasser, 2000). Unfortunately, the snapshots that are a part of these children's typical experiences may lead to their own victimization or abuse of others (Pepler, Catallo, & Moore, 2000). Negative coping mechanisms develop in children who witness violence or who are maltreated in order to deal with intense emotions accompanied by a lack of appropriate cognitive perspective for the behaviors of their caretaking adult providers (Sikes & Hays, 2010).

Key to successful relationship outcomes is the degree to which the basic needs are satisfied so that early life events ideally enhance the ability for individuals to assume satisfying adult roles in relationships (Godbout et al., 2009). However, all too often domestic violence is the "elephant in the room" or the obvious topic that everyone is aware of, but which is not discussed because the topic is uncomfortable. What often occurs in adulthood is confusion separating the effects of childhood trauma from the normal developmental processes (Bedi & Goddard, 2007). Most importantly, poor outcomes during childhood, such as posttraumatic stress reactions and mood disorders, or what children's teachers call "externalizing behaviors," influence the effectiveness of relationship skills throughout life (Bedi & Goddard, 2007).

The couple that is the subject of the following case study is Ava and Anderson. They met in college and were married when they were in their early 20s. This couple had their first child prior to setting the wedding date and at the time of our meetings were the parents of three additional children. They are college-educated African Americans. Anderson's mother was physically abused and killed by one of her abusers. Anderson was also abused by the mother's boyfriend/killer. His mother believed in and practiced corporal punishment so, consequently, Anderson believes corporal punishment is appropriate for his and Ava's children. Ava's mother was a survivor of domestic violence and Ava experienced vicarious trauma due to her mother's abuse. She also experienced some punishment in her childhood (i.e., spanking, harsh words, etc.). Interestingly, Ava and Anderson rarely talked about the similarities of their pasts.

CASE STUDY

Ava called the counseling office to make the initial counseling appointment. She responded to questions regarding the nature of the problem by saying that she hated her husband and everything about him. After she calmed down, the family therapist asked for clarification regarding this statement. An appointment was made and Anderson agreed to come with Ava.

Session 1

In the initial visit, I asked Ava and Anderson to tell me about some of the challenges they were experiencing. I explained that I am a "doing" (R. E. Wubbolding, personal communication, July 4, 2010) kind of therapist and asked them if this was satisfactory to them. When they answered in the affirmative, I told them this would mean they would have to spend time doing some things differently between sessions and then report back on how they were affected. I also asked Ava to talk more about what she meant when she said she hated her husband. Ava looked at me with an expression of disgust on her face and said, "What do I mean? I hate him!" I responded, "Hate is a strong word." She said, "But, I mean it."

I could tell by the expression on her face that she was serious. While Ava spoke of hating him, Anderson did not speak or show any emotions. It became clear to me that this relationship was in serious trouble and that it might be helpful to explore how the relationship had deteriorated. I wondered what frustrating unmet need could be causing this woman to loathe her husband. The couple did not look at each other once during the first few minutes of this session. I allowed Ava to continue talking. I asked her how she had gotten to this place and she responded,

> It's easy when it went from me looking at him and thinking, "I love him so much" to "Why am I even looking at you?" It was more or less that he went from being so steady and constant to someone who wasn't anymore.

I asked Anderson if there was something he wanted to say in response to his wife's present assertion. He dropped his head and replied,

> I've done the right thing for so long that I just felt like I needed to do something for me. She was always responsible and she always did what was right. Look what happened to her. Without achieving any of her dreams, look what happened to her.

I asked Anderson, "Who is the 'her' you are referring to?" Ava did not give him a chance to respond, saying,

> It was his mother. She has been dead for 15 years. He has not grieved her death, and he has not let her go. Initially, I felt sorry for him. If this would bring him closer to her, I was all for it. But, when we started getting all those [late bill] notices, I was fed up with this! I just want things to be the way we planned.

Anderson sat there quietly for a few seconds and then he said, "I thought I had it under control. It just stopped feeling good to pay bills." I commented,

I am sure you love your mother and want to honor her memory, but how does what you are doing help your current situation? How can we help you honor her memory in such a way as to allow you to do what you need to do in the present?

The couple was silent and I questioned Anderson further, "Would you be willing to think about how this might be keeping you from moving forward? Would you think about this with me?" Using an analogy, I asked Anderson, "If you tie your little toe to the door over there with a string that is about 3 feet long, how far would you get when you try to leave the space?" (I have used this example in the past to help clients think about how the past is related to the present.) "Three feet," Anderson replied. I went on:

So, could you think about how this might be similar to being tied to the past? You're not going to get very far if you are tied to the door. You won't get very far if you stay focused on the past. It's also like looking in the rear view mirror while you are trying to drive forward. What happens?

Anderson said, "You don't get very far, except maybe in an accident." I added, "Do you see where I'm going with this?" He said, "Yeah, I think so. But, I'm going to have to think about it for a while." I told him that that was okay and he could think about it as long as he needed to. I was sure we would talk about it more in the future.

Anderson had been living with the past for a long time and was not expected to change his thinking in one session. However, it was important that both of them do something different very quickly. Anderson needed to loosen his grip on the past and Ava needed to realize her expressions of hatred toward Anderson were unproductive. It was also important to determine what the couple's purpose was for seeing a counselor. I asked, "Are you here for couples counseling or divorce counseling?" Ava responded, "Frankly, we wouldn't be here paying you if we were planning for a divorce." Anderson looked at her, then at me, and said, "That's why I'm here, too." They were asked if they both thought the marriage was worth fighting for. They both replied affirmatively.

Need Strength Profile Assessment The couple individually completed a genogram, clinical interview, and the choice theory need strength profile rating scale. Their scores are described in Table 9.1.

Table 9.1 Ava and Anderson's Scores on the Choice Theory Need Strength Profile Rating Scale

Needs	Love and Belonging	Power	Freedom	Fun	Survival
Anderson					
Need strength	8	8	6	7	6
Need satisfaction	6	7	7	6	7
Ava					
Need strength	8	9	10	8	5
Need satisfaction	5	8	6	4	2

With the exception of the "freedom" need, Ava and Anderson were within one point of each other on the need strength category. According to Glasser and Glasser (2000), these similarities are good and when there is a significant difference, the couple can be instructed to learn to work with or through this situation.

In terms of the need for love and belonging, Anderson and Ava had the same need strength in this area and both indicated this need was being met at a level greater than the halfway mark. Though the match between a want and the level of perceived need satisfaction was not as low as it could have been, the total behaviors that were observed were distinctly communicating a lack of commitment to the relationship. As Wubbolding and Brickell (2005) noted, communication is a purposeful behavior to impact the external world. Anderson's less frequent verbalizations instilled a lack of interest in Ava's perception, so, in order to get his attention, Ava chose communicating with what Glasser called the "deadly habits" (Glasser & Glasser, 2000).

For the "power" need, Ava had a slightly higher rating than Anderson, but this did not appear to be a concern in this couple's perception. Ava's satisfaction of the power need appeared to be greater than Anderson's, as well. Ava used external control psychology unwittingly in nagging Anderson to behave in ways that were satisfying to her expectations (Glasser & Glasser, 2000). What became apparent was that Anderson's description of his feelings related to his mother's life and death was, in fact, his method of coping with his power need to influence his current actions. For example, he had recently begun to spend money unwisely. According to Anderson, his mother was always doing the right thing, paying the bills, and then she had died without anything of substance to show for it. His attempt to avoid what he perceived to be doing the right things like his mother in the past was creating maladaptive behaviors in his relationship in the present. His role in the marriage had evolved into that of the money manager, a position of control over the family.

For the need for freedom, Ava had a significantly higher need than did Anderson. This need difference of four points could create a problem for the couple. Ava's high freedom need did not appear to create significant dissonance in the family at this time. She did talk about wanting to get away from family, children, and husband; sometimes she just needed to be alone or out of the house. She took care of this by getting in the car and going for a long ride. However, having several similar need strengths should work in the couple's favor as it relates to relationship longevity.

The couple's scores, again, were within one point of each other on their need strength for fun, but Ava appeared to be less satisfied than Anderson in this area. Though one might extrapolate that this need is less significant than others for marital satisfaction, the fact that the couple did not enjoy being together and enjoying each other's company spoke volumes about the current condition of the marriage. Glasser and Glasser (2000) state that fun is a need built into the genetic makeup of almost all animals. Fun is also how humans learn, which also contributes to survival advantages for individuals or survival of the relationship (Glasser & Glasser, 2000).

Finally, for the survival need, the couple scored within one point of each other, but the difference in their need satisfaction scores also was a significant problem for their relationship. Ava's self-reported rating of a 2 for need satisfaction is a significant indicator of a lack of safety. On the surface, Anderson's need satisfaction rates were higher than his need strength, which seemed to signify his perception that he had all he needed in terms of basic physical needs and gratification. His apparent security was conjectured to be representative of safety in the relationship continuation for living support if nothing else.

The first goal was determined to be assisting this couple to find a way to talk with each other without causing any additional pain. The benefits of learning choice theory and then practicing it with each other were apparent for this couple. A review of each client's genogram suggested that Anderson was not in contact with and did not have many family members, whereas Ava had a large extended family. One of the reasons this couple had been able to maintain their relationship was that Anderson began to connect with her large family early in their relationship. The connections he had nurtured over the years with Ava's family members represented close ties that he lacked in his upbringing.

Caring and Deadly Habits   At this time, it would be beneficial to introduce the importance of using Dr. Glasser's caring habits. Glasser (2000) identified seven caring habits that lead to satisfying and fulfilling relationships: supporting, encouraging, listening, accepting, trusting, respecting, and negotiating differences with people we care about, which would nourish relationships with others. In contrast, Dr. Glasser describes the seven deadly habits: criticizing, blaming, complaining, nagging, threatening, punishing, and bribing or rewarding to control. It was important to teach this couple to change the way they responded to each other and to recognize the inappropriate way they communicated with each other. Therefore, it was appropriate to suggest they use Dr. Glasser's caring habits more and the deadly habits less.

Toward the end of the first session, the couple was asked to spend some time over the next few days thinking about how often they used the deadly habits. They were asked to keep a log of the number of times they found themselves displaying the seven deadly habits. They agreed to do this for 7 days and then return to the office.

Session 2

At the second meeting, Ava declared she did not realize how much or how often she used the deadly habits discussed the week before. Anderson was less excited about the whole idea but did say that he noticed how much she used them. He admitted using the deadly habits quite a bit himself. Anderson said, "I can see why you call them disconnecting habits because every time she starts to nag I want to disappear."

Not surprisingly, talk of Anderson's mother reentered the second session. He talked about what he might do to stop himself from a desire to go back to the day she was killed and rescue her. Though the past may impact events in the present, dwelling on historical situations does not create an environment in which potential for change can be explored (Wubbolding, 2011). However, when there is a history of trauma or abuse, it is appropriate to gather information about the context of the client's experiences to clarify the issues (Wubbolding & Brickell, 2008). The knowledge that Anderson could not do anything as a young child to prevent his mother's death did not change the fact that he continued to think about her almost daily and about how his life could have been different if he had saved her.

Although Anderson was certainly not at fault for his mother's murder, it was suggested that he consider forgiving himself for what he perceived as a failure on his part. At that point, he mentioned his efforts to forgive other family members for his mother's demise for not helping her out of the abusive relationship. Anderson's whole presence changed in that session, as did Ava's as she heard the depth of the emotions Anderson had been dealing with but had not shared with her.

The definitions of guilt, blame, and responsibility were discussed, and the session focused on Anderson's ability to embrace the reality that he had no guilt. Through the language of choice theory, it was stated that Anderson could choose to relieve himself of any guilt surrounding his mother's death. It was not uncommon for children to be torn between a desire to help or rescue the victim and the need to keep a family secret (Bedi & Goddard, 2007). Children respond to conflict with various forms of distress and often are confused with feelings of inappropriate responsibility for persons and events—hence, the attempts to save the victim of assault. He said he would work on forgiving himself.

At the end of the session, Anderson was encouraged to exercise self-forgiveness, and both Ava and Anderson were assigned to continue their practice of using caring habits and minimizing the use of deadly habits. The session ended with the couple focusing their attention on how they could use choice theory tools in their relationship. Ava said she was sure that she was already using most of the caring habits, but would try to use them more often in the next week. Anderson said he would also try to use the more positive language that Glasser (1998) proposed in the caring habits.

Session 3

At the next visit, Ava was in a somewhat better mood. She stated that she had practiced using more of the caring habits talked about the week before and she continued to try to use the deadly habits less and less. She felt there was some change in the way she and Anderson were communicating and getting along. She also reported making an effort to listen to him when he talked to her and encouraging him in some of the things he wanted to do. Anderson smiled when he heard her say these things. He reported that he knew something was different but he was not sure what it was. Choice theory appeared to be working for this couple. An insight occurred to Ava about how the tools of choice theory might be a way for them to be more successful in their marriage.

Sessions 4 and 5

For the next two sessions, the primary focus was on continuing to support the couple in replacing the deadly habits with which they had grown so familiar. Reports included fun and enjoyment as a family and as a couple. Glasser and Glasser (2000) state the importance of the survival advantage of learning, which is closely tied to the ability to have fun. Fun also is understood to increase the opportunity to experience a sense of inner joy (Wubbolding, 2011).

Session 6

At the sixth visit, Ava and Anderson came in and sat down close to each other. They were both smiling and talking about a song they heard in the car on the way to the office. This behavior change was significant because Ava would not even look at Anderson when they began counseling. Anderson was alert and seemed more eager to participate. I asked them how things were going for them at home and they responded almost in unison, "Great." When I asked each of them to expand on this comment, Ava said, "It is no longer a chore to get the trash taken out," and Anderson said, "I don't know what she has been doing but things are very different for us." Ava was leading and he was following, appropriately.

In the same visit, Anderson said he never thought about forgiving himself for his mother's death. He knew cognitively he had nothing to do with it, but because he was the last one of the children to see her alive, he had always thought maybe he could have said or done something that morning to change the tragedy. He reported feeling better about letting go of the past and living fully in the present with his family.

Session 7

At the last session, Ava and Anderson completed the need strength profile rating scale again, and the comparison of ratings was discussed (Table 9.2). It was evident that all of Anderson's ratings increased over the course of all sessions. His interpretation was that the increase in caring habits, or what he first described as loving behaviors, diminished his feelings of fear, insecurity, and jealousy (Mickel & Hall, 2008).

Table 9.2 Need Strength Profile Rating Scale Comparisons

Needs	Love and Belonging	Power	Freedom	Fun	Survival
Anderson					
Need satisfaction—first round	6	7	7	6	7
Need satisfaction—second round	7	7	8	8	9
Ava					
Need satisfaction—first round	5	8	6	4	2
Need satisfaction—second round	9	6	7	7	7

On the other hand, Ava had one rating drop in the power need. It was important to get a grasp on how she perceived such a change. Her response was that she could let go of some control in the relationship and put her energy into more effective and satisfying behaviors, such as attempts to have more fun. Need fulfillment in a significant relationship may be expressed through a balance of attention to all domains of total behavior, such as her thoughts about what was important to both her and Anderson, or the actions that she was committed to in using caring habits in their communication (Mickel & Hall, 2009). The more balance in the need satisfaction for both partners, the more reflective of need fulfillment is in the relationship. Finally, the attention to each other's quality world pictures enhanced the couple's development of a potential future for the couple and not just two individuals.

Ava and Anderson had been using the caring habits for several weeks and it appeared to be a good time to introduce the idea of the solving circle. They were told that it would help when they found themselves having some difficulty with a decision. The solving circle is a concept developed by Glasser and Glasser (2000). Ava and Anderson were asked to visualize themselves in a circle. While in the circle they were to think of the marriage as of primary importance and focus. No matter what the problem was, they were to visualize the marriage as more important than anything else. Ava and Anderson were instructed to step into the circle whenever they needed to make an important decision. When doing so, they also were to think about whether or not an action was helpful or harmful to the marriage. If it was going to hurt the marriage, then the choice was about whether the action was something they really wanted to do.

Ava and Anderson practiced this in session. I reinforced for them that our session was a good place for them to start practicing the skills they were learning to use on their own. They concluded counseling with the understanding that they could return at any time they felt they needed to. I also asked if it would be all right if I contacted them within 6 months to see how things were going, and they agreed.

CONCLUSION

Ava and Anderson changed the way they communicated with each other. They learned to use the caring habits and gradually used fewer and fewer of the deadly habits. They demonstrated an understanding of the solving circle. Ava was more receptive to the idea of therapy and changing her behaviors than Anderson. According to Ava, she changed the way she responded to Anderson, using fewer of the deadly habits. At one point, Anderson stated he had not used all of the things we talked about but something was different in their marriage. He was receptive to Ava's changes.

Initially, Ava and Anderson were congratulated for coming to therapy. This was a major accomplishment for many reasons. They were African American, young, and experiencing some significant marital problems. They had not given up on their marriage and were looking for support from a nontraditional source: therapy. This African American couple had learned to use choice theory/reality therapy to enhance their marriage and increased the possibility of staying together. When we think about the significance of the national statistics—that 33% of all marriages and 47% of African American marriages end in divorce (Bramlett & Mosher, 2001)—it is important that therapists practice methods that increase effectiveness in keeping couples together.

REFERENCES

Bedi, G., & Goddard, C. (2007). Intimate partner violence: What are the impacts on children? *Australian Psychologist, 42*(1), 66–77.

Bramlett, M. D., & Mosher, W. D. (2001). *First marriage dissolution, divorce, and remarriage: United States.* Centers for Disease Control and Prevention (CDC Publication No. 323).

Edleson, J. L. (1999). The overlap between child maltreatment and woman battering. *Violence Against Women, 5,* 134–154.

Glasser, W. (1998). *Choice theory: A new psychology of personal freedom.* New York, NY: HarperCollins.

Glasser, W. (2000). *Counseling with choice theory: The new reality therapy.* New York, NY: HarperCollins.

Glasser, W., & Glasser, C. (2000). *Getting together and staying together: Solving the mystery of marriage.* New York, NY: HarperCollins.

Godbout, N., Dutton, D., Lussier, Y., & Sabourin, S. (2009). Early exposure to violence, domestic violence, attachment representations, and marital adjustment. *Personal Relationships, 16,* 365–384.

Mickel, E., & Hall, C. (2008). Family therapy in transition: Love is a healing behavior. *International Journal of Reality Therapy, 25*(2), 32–35.

Mickel, E., & Hall, C. (2009). Choosing to love: Basic needs and significant relationships. *International Journal of Reality Therapy, 28*(2), 24–27.

Mickel, E., & Wilson, S. I. (2004). Family therapy in transition: Connecting African centered family therapy with a multisystems approach. *International Journal of Reality Therapy, 23*(2), 31–35.

National Coalition Against Domestic Violence (2007). Domestic violence facts. http://www.ncadv.org/files/DomesticViolenceFactSheet(National).pdf

Pepler, D. J., Catallo, R., & Moore, T. E. (2000). Consider the children: Research informing interventions for children exposed to domestic violence. *Journal of Aggression, Maltreatment and Trauma, 3*(1), 37–57.

Sikes, A., & Hays, D. G. (2010). The developmental impact of child abuse on adulthood: Implications for counselors. *Adultspan Journal, 9*(1), 26–35.

Wubbolding, R. E. (2000). *Reality therapy for the 21st century.* New York, NY: Routledge.

Wubbolding, R. E. (2011). *Reality therapy: Theories of psychotherapy series.* Washington, DC: American Psychological Association.

Wubbolding, R. E., & Brickell, J. (2005). Purpose of behavior: Language and levels of commitment. *International Journal of Reality Therapy, 25*(1), 39–41.

Wubbolding, R. E., & Brickell, J. (2008). Frequently asked questions and not so brief answers. Part II. *International Journal of Reality Therapy, 27*(2), 46–49.

10

COUNSELING INTERFAITH COUPLES

Neresa B. Minatrea and Jill D. Duba

INTRODUCTION

Most dating services, online or otherwise, try to "match" people based upon similar interests, values, likes and dislikes, and short- and long-term goals. Information used to make successful matches includes culture or racial identity, socioeconomic status (salary) and religious and/or religious denomination affiliation, interests, values, and education. It is not surprising that most people looking for a potential partner want him or her to "match" at least one or more of the criteria mentioned. For the majority of Americans, marrying someone who is demographically similar seems to be the norm (Gardyn, 2002). On the other hand, there are considerable cultural, social, and religious pressures that actually dissuade individuals from coupling within someone who is much different from them (Lara & Duba Onedera, 2008).

Some couples do not fit the "norm." That is, many people couple with others who are socially, culturally, and religiously different. In the case of religious difference, for example, the frequency of inter-religious marriages (a marriage between persons of differing religions) is on the rise. In fact, one partner in every 23% of Catholic marriages, 27% of Jewish marriages, 21% of Muslim marriages, and 33% of Protestant marriages does not identify with that particular religious faith (Kosmin, Mayer, & Keysar, 2001). The number of interfaith marriages is also on the rise.

Interfaith marriages are marriages between persons of the same religious affiliation (e.g., Christian) with slightly different practices (e.g., Baptist and Methodist practices).

INTERFAITH AND INTER-RELIGIOUS COUPLES: WATERS TO NEGOTIATE

Many inter-religious and interfaith couples face particular struggles and challenges. Couples who share the same degree of religiosity will experience a shared meaning about what God expects from them as married individuals, as well as what each spouse can likely expect from the other (Duba & Nims, in press). For interfaith and inter-religious couples, shared meaning will come from negotiations, concessions, and compatibility. According to Lehrer and Chiswick (1993), a couple's compatibility "dominates any adverse effects of differences in the religious background" (p. 400). However, when religious perspectives differ, so also may views and expectations. Interfaith couples must be able to negotiate and work together on issues related to communication, forgiveness, equity, togetherness, intimacy, love, sexual intimacy, and commitment (Bryant, Conger, & Meehan, 2001; Fincham & Beach, 2002; Frame, 2004; Weigel & Ballard-Reisch, 1999). Finding a "shared meaning" that fits for both partners regarding any of these issues is an essential step in clarifying and negotiating what they believe that God expects from them as a couple and what they expect from one another.

Negotiations

For many religious persons, participating in activities related to religion is not only a must but also a preference (Duba, 2009). When couples share different religious perspectives and practices, this can threaten time together. Some couples must negotiate if or how to celebrate different religious holiday functions together (i.e., Christmas, Hanukah). Such decisions may be emotionally trying. For example, how will a Jewish partner participate in Easter-related activities without feeling that his or her Jewish identity is threatened? How does this affect the relationship? How does a Catholic mother feel about acquiescing to her Baptist husband's request to baptize the baby when he or she is grown? In some ways, making such decisions is not about compromising with one's partner, but rather about coming to terms with how one makes meaning of events, traditions, and customs. Further, the individual partner must negotiate personal beliefs and experiences at the intrapersonal level.

Expression of sexuality is often determined by a religious person's faith and/or related doctrine. For example, a traditional Muslim husband may assume that his sexual gratification is important and his wife's response to his sexual need is equally important (Duba & Nims, in press). If his wife is a liberal Christian, negotiating about sexual practice and sexual "equity" may not be such an easy task. Even interfaith couples may struggle with making sexual conciliations. A Catholic partner, for example, may feel strongly about not using contraception. This partner may rely on the Church's teachings that every act of marital intercourse should be one that is open to new life (Duba, 2008). The Protestant partner, on the other hand, will likely believe that there is not a biblical condemnation of contraception (Zink, 2008). Contraception and sexual gratification are only two possible areas on a wide spectrum of various religion-based differences regarding sexuality.

The degree of perceived equity may differ in couples who share varying religious viewpoints. For example, the patriarchal nature of the Muslim family, as well as various references in the Quran, provides married men and women with the expected ways of contributing to the marriage, as well as to the livelihood of the spouse (Duba & Nims, in press). Will a Muslim husband be able to meet his Christian wife halfway, particularly if the wife does not subscribe to Muslim-based expectations of wives? Alternatively, will an interfaith, Protestant/Catholic couple be able to compromise, be respectful of each other's differences, and coexist in the same family unit? Perhaps the Protestant husband's father was a preacher and the family subscribed to traditional gender roles. The Catholic wife, on the other hand, came from a family where there was less emphasis on traditional gender roles. This couple may need to discuss how their expectations for gender roles are embedded within the context of their faith, as well as what they are willing to compromise for the betterment of the relationship.

CASE STUDY

A neurologist referred Keith, a 36-year-old man, for seizures resulting from taking too much tramadol (Ultram). This had been Keith's second visit to the emergency room for overmedicating; consequently, his physician insisted that Keith pursue counseling for his addiction to tramadol and prescribed no driving for 6 months. Clients come to counseling through a myriad of pathways that sometimes lead to a different set of concerns. In Keith's case, he quickly focused upon his misuse of pain medication and agreed to abstain from taking pain medications; however, a systemic concern quickly arose surrounding his interfaith marriage.

The couple, Keith and Sue, indicated that their romance started while they were in college. At that time, both envisioned having home, family, career, and happy family holidays. Like many dating couples, they discussed family, parenting, and their different religious practices. Since neither Sue nor Keith was "devout" in religious practices, they had decided that their differing interfaith practices would not be a problem. As a child, Keith attended a local Methodist Sunday school and worship services with his parents, younger brother, and sister. Once Keith left home for college, he did not "attend church as he should." Unlike Keith, Sue described attending Catholic Mass with her two older brothers and parents throughout her childhood and into her college years. During Keith and Sue's dating years, Sue attended Mass with her extended family and diligently practiced her faith. Keith stated that early in the relationship, their different interfaith beliefs and practices had presented some distress and friction regarding holiday practices, sexual intimacy, marriage and baptism ceremonies, and beliefs about the afterlife. The couple had negotiated and discussed the marriage ceremony with their families and Sue's priest and they agreed to a Catholic ceremony to seal the marriage vows.

Following 5 years of marriage and experiencing disillusionment with their yearlong *in-vitro fertilization*, Keith and Sue adopted a baby, Sam, from Asia. Subsequent to adopting their baby boy, Keith reported increased feelings of sadness and melancholy, and thoughts surrounding lack of motivation, stating, "I was in a rut." He described boredom with going to work, coming home, sleeping, cleaning house, and then starting the routine all over the next week. During this time, Keith reported that he began taking tramadol due to a back injury and continued the medication, saying that it helped him "to escape."

Sue reported feeling perfectly satisfied, living her dream of attending to Sam's daily needs and running household errands, a lifestyle similar to the model of her own mother during Sue's childhood. Sue conveyed that she was happy living her dream, caring for a baby and for a husband she loved. She stated, "I'm tired after chasing Sam all day," and she expressed no desire to participate in social activities except a few family events and Mass. At the point of counseling intervention, this family was busy with routine daily life events and stressors. However, taking advantage of quality time with each other or attending social or religious activities as a family appeared to be at a minimum.

Case Conceptualization

From a choice theory/reality therapy (CT/RT) perspective, Sue and Keith exhibited several similarities in their backgrounds and in their individual quality worlds—that is, their ideal pictures of what they wanted in their lives (Glasser, 1998). They appeared to have similar values and interests regarding their lifestyle (middle-class economic status), quality family time, commitment to marriage, Christian childhood guidance, and education. In contrast, variations in their quality world pictures of what was important to them emerged, including how they viewed extended family involvement, Christian practices (Catholic versus Methodist), career aspirations, and leisure activities. These distinctions appeared to be manifesting into interpersonal and intrapersonal conflicts within the family.

Specifically, Keith's high needs for power, fun, and belonging with friends seemed to conflict with Sue's high need for freedom and low needs for power, fun, and belonging. This led to many frequent debates about participating in activities with friends and family, housekeeping, childcare, and religious practices. For instance, Sue remained at home providing for Sam and did not engage in many outside activities with friends, but Keith desired to participate in activities with neighbors and friends two to three times a week and especially on weekends. The couple's different interfaith practices were becoming more distressing regarding holiday rituals, Sam's schooling, and family activities. Organizing and completing household chores and finances presented additional challenges for this couple due to their different perceptions about how these jobs should be handled. Consequently, the family continued each week maintaining the same behaviors and routines, yet feeling very unsatisfied with their perceptions of how they were meeting their needs. When Keith entered therapy due to the seizures from tramadol, his neurologist restricted his driving for 6 months, prescribed no new pain medication, and required him to participate in counseling.

The Counseling Process

The therapist used CT/RT (Glasser, 1965, 1984, 1995, 1998, 2000a, 2000b, 2003; Wubbolding, 1986, 1991, 2000, 2011) as the counseling theory and techniques to ascertain Keith's and Sue's individual needs, wants, and quality world pictures, and to identify what they desired as a couple. Because of the directives from Keith's neurologist, during the early sessions, the life-threatening issue of abusing tramadol becomes the focal point. Keith focused on abstaining from pain medications by creating several total behavior charts (Wubbolding, 1986, 1991, 2000), which helped him identify his actions, thoughts, feelings, and physiology. Completing total behavior charts provided an epiphany for Keith: "I'm spending most nights and most of the weekends doing the same thing: sitting in my chair watching Sam play while watching TV. I'm bored." He further identified his thoughts: "I'm tired" and "I'm lonely." Associated with his inactivity and thoughts of boredom, his backaches and pains appeared to increase; he felt sad, gloomy, and discouraged.

After identifying each current behavior within the total behavior chart, Keith evaluated his behaviors and goals and then made a new chart outlining the way he *wanted* things to be in his life. In the course of these exercises, Keith indicated a desire to participate in church activities, thus fulfilling several of his basic needs (belonging, power, and fun). After each session, he defended himself against old behaviors with his rolled-up chart, stating, "I have a mission plan now." Part of his "mission plan" included his wife's involvement in family and individual sessions.

Family Systems Work   During the early counseling experience, the couple described their family dynamics. A three-generation genogram was drafted (Duba, Graham, Britzman, & Minatrea, 2009). The couple experienced transgenerational learning as a result of identifying patterns, disorders, illness, and beliefs within their extended families. They also recognized patterns in their current behaviors. "Sue's family are always doing things together and coming over to the house, while my family does very little together throughout the year." They further pointed out that Sue's family attended numerous events, such as Catholic Mass, christening of babies, birthday celebrations, and other religious holiday celebrations. Another pattern Keith recognized involved display of affection: "My family does not show a lot of emotions, whereas Sue's family are always hugging and saying 'I love you.'" Although the couple discussed numerous patterns, the following appeared relevant to Keith's recovery. Keith's family members used a great deal of prescription drugs and they gave each other medication.

The genogram exercise supplies an image of the couple's quality world and understanding about their current behaviors. Subsequent to this exercise, Keith announced his new want of "attending church as a family" to be a part of his total behavior plan. For several reasons, the couple chose Catholicism over the Methodist faith:

> The Catholic Church appeared to be a strong influencing part of Sue's family.
> Sue's family was very involved in their life.
> The couple wanted Sam to attend a private Catholic school.

Keith started his new mission of discovering the requirements, training, and rituals to convert to the Catholic faith.

Experiential Work To help the couple discover and share their quality world pictures and basic needs, the therapist asked them to engage in an exercise in which they drew a ship in a storm (Oklander, 1988, 2006). In this activity, the clients draw their version of a ship in a storm with a lighthouse. The therapist uses the drawing to facilitate questions evaluating the couple's perceptions and the behaviors chosen to meet their needs. The therapist also discovers how they navigate their current life stressors.

Sue and Keith noticed metaphors (Wubbolding, 1991) and generalizations applicable to their current life. During this exercise, Keith said, "My ship does not have a harbor or safe place to go. ... I don't have anything to navigate my ship." Sue acknowledged, "I am in the hull of the ship where it is safe with Sam. ... I am at the mercy of the storm." This exercise helped in their understanding of each other's quality world pictures and the current behaviors they employed to meet their wants and needs. Keith's picture expressed a high need for power and a desire to be in charge, whereas Sue desired safety and someone else to be "steering the ship." This led to evaluating their current behaviors and recognizing new behaviors, moving them toward meeting their wants and needs in a healthier manner as a family.

Needs Strength Profile Taking a needs strength profile (Glasser, 1995) can help a couple examine their perceptions, quality world pictures, and basic needs. This profile identified the similarities and inconsistencies between Keith's and Sue's perceptions of their needs. The needs are for love and belonging, power, freedom, fun, and survival (Glasser, 1998). In this activity, individuals are asked to score from 1 to 10 the strength or intensity for each of their five basic needs, with 1 being the lowest and 10 being the strongest.

Starting with the need for power, Keith reported a higher score (9) as exhibited in his striving for leadership roles, creating and accomplishing projects within his career, spiritual, and leisure activities. Keith's high need for power even manifested itself in the Catholic conversion process by his taking control of the time he wanted to attend training and by choosing where he wanted to study the faith. Sue met her need for power through "being a good mom" and indicated a very low need for power (3 out of 10); additionally, she expressed no desire to return to graduate school, pursue career goals, and, within their relationship, was OK with Keith making most of the financial decisions and bringing in their income.

The difference between the intensity in their need for power (9 and 3, respectively) provided insight into some of their relationship struggles.

Keith wanted power through achievements at work, home, and with his leisure activities, while Sue was OK with achieving her need for power through activities surrounding her child. Frequently, Keith wanted to impose his high need for "family goals" upon Sue.

Next, their differing appraisal for love and belonging appeared indicative of some other struggles. Sue had an intensity score of 8 for family and only 3 for friends, whereas Keith's was an 8 for friends and 3 for family. The couple described their behaviors in meeting this need in very different ways. For example, Sue was content to spending multiple days just with her son and husband, while Keith wanted to participate in activities involving friends numerous times during a typical week.

Regarding the need for fun, Keith's high need for power and belonging with friends dovetailed with his high need for fun at an intensity of 9, which involved learning new things, being with friends three to four times a week, and participating in sports activities. This conflicted with Sue's idea of fun at a 3, which involved staying at home, watching Sam play, and viewing movies. "Keith doesn't understand that I am so worn out after a day with Sam, I just want to stay home," she said.

Sue rated her basic need of freedom at 8 out of 10, higher than Keith's score of 4. This is indicative of Keith's comments that "the day needs to be planned out or nothing gets accomplished." Sue enjoyed not being bound by time or structure except when it related to Sam's eating and sleeping times.

Lastly, the survival need scores appeared comparable for the couple; Sue and Keith perceived this at a 7 and 8, respectively. They valued paying bills on a timely basis, having freedom from debt, building a substantial savings account, and planning for Sam's future.

From this needs profile, the couple identified several goals using the SAMI^{2}C^3 (simple, attainable, measurable, immediate, consistent) over P model (Wubbolding, 1986, 2000, 2011). Both agreed, for the good of the relationship, to attend Mass, complete a Catholic couples training course, go out with friends once a week, go on a date two times a month, and go out individually with friends once a week. Throughout the counseling process, Keith and Sue revisited this needs profile and their goals to evaluate their progress.

Myers-Briggs Type Indicator and Reality Therapy　A last activity in this case involved integrating the Myers-Briggs type indicator (MBTI) and reality therapy (Minatrea & Ophelan, 2000). This communication activity increased Sue and Keith's acceptance of each other's preferences and use of connecting habits (Glasser, 1995). The MBTI (Hirsh & Kummerow, 1989; Keirsey & Bates, 1978; Kroeger & Thuesen, 1988;

Myers & McCaulley, 1990; Tieger & Barron-Tieger, 1992) assesses an individual's personality preference using 16 types. The combination of four basic dichotomous preferences identifies 16 personality preferences (Myers & McCaulley, 1990).

The first set entails the way one draws energy and makes sense of the world: extroversion (E) and introversion (I). Next, sensing (S) and intuition (N) refer to the way one assimilates information using senses or insight. Individuals who favor sensing tend to gather facts and information prior to making decisions, while innovative and intuitive persons frequently rely on their hunches and then gather information to support their decision making. The third set of polar preferences, thinking (T) and feeling (F), encompasses the way one makes decisions—either from objectivity, using rules and logical connections, or from a personal point of view, emphasizing values and the feelings of others. The fourth set describes how individuals may prefer their outer behavior to be. A judging (J) type indicates individuals who are organized, purposeful, and decisive. In contrast, those who identify a trait of perceiving (P) may exhibit behaviors involving spontaneity, creativity, keeping their options open, and suspending judgment until they have gathered more information. The combination of these four sets of preferences provides insight into how clients interact with the world and how they attempt to meet their five basic needs (Minatrea & Ophelan, 2000).

Using the MBTI to evaluate how this couple met their five basic needs provided an evaluation tool in understanding their quality world and facilitated connecting habits. Keith's personality preference scores indicated ESTJ (extrovert, sensing, thinking, judging), whereas Sue's type reflected INFP (introvert, intuitive, feeling, perceiving). Briefly, this meant that this couple satisfied their needs through their personality preferences; therefore, meeting their five basic needs came about in very different ways. For example, Keith oriented himself to the world and renewed his energy extrovertly by talking about things vocally and being around friends and family. Meanwhile, Sue was an introvert who preferred staying at home, reading a good book, watching a movie, and being with close family.

Considering their differences in thinking and feeling, and in judging versus perceiving, Keith and Sue discussed their preferences. Keith talked about his inclination for judging and Sue compared her behaviors, identifying with perception in this category. Keith indicated, "I like everything to have a place and it drives me nuts when all the flat surfaces in the house have piles of stuff." This comment was indicative of how this personality preference wanted structure and valued time (Tieger & Barron-Tieger, 1992).

The discussion moved to Sue's difficulties in being ready to leave the house because she could not find things. A recent argument involving arriving on time to Sunday Mass prompted Sue to counter, "Well, you always want to be early and that's why I am not ready to go to Mass with you." The tendency for judging or perception was evidence of how Sue met her high need for freedom and Keith's behaviors gratified his need for power through organization, prioritizing, and attention to detail.

Lastly, the two middle dimensions in the MBTI are sensing and thinking (ST) for Keith and intuitive and feeling (NF) for Sue. These dichotomous sets describe how individuals take in new information and make decisions (Tieger & Barron-Tieger, 1992). Keith explained the need to make lists, research decisions, and apply the rules equally. Meanwhile, Sue at times just instinctively knew the best decision and she considered family's or friends' feelings when making tough choices (Kroeger & Thuesen, 1988).

These two preferences manifested themselves very differently with the couple's decision to choose to practice Catholicism and to educate Sam in a Catholic school. Keith reported researching various religious practices, talking to the leaders of the Methodist and Catholic churches, and visiting the Catholic school. In contrast, Sue stated, "I just knew that was how I was raised and that is my want for Sam." These two preferences influenced the couple's decisions about attendance at family events and, recently, about childcare. "If we don't let Mom [Sue's mom] keep Sam, her feelings will be hurt when we go out on our Friday night date." The interweave between the aforementioned needs strength profile (Glasser, 1995) and the MBTI (Myers & McCaulley, 1990) personality preference enhanced Sue and Keith's understanding of each other's quality worlds, thereby improving their connecting behaviors (Glasser, 1995; Minatrea & O'Phelan, 2000).

Summary of the Counseling Process

The aforementioned highlighted family sessions, which initially focused on Keith's overuse of pain medication, moved to systemic concerns involving disconnecting habits (Glasser, 1995) and the identification of behaviors that were not meeting the couple's or individual's needs (Wubbolding, 1986, 2000, 2011). Numerous CT/RT techniques and strategies facilitated the couple's knowledge about their perceptions, wants, quality worlds, and connecting habits. Throughout the time the couple engaged in counseling, they utilized the total behavior chart to evaluate their progress toward family and individual goals.

The family attended family and individual counseling for 2 years—initially, weekly, and then decreasing to once a month. Keith experienced two brief relapses during the 2 years, both after 60 days of recovery. Keith completed his Catholicism classes with a ceremony. Keith and Sue found they could actively meet their needs for belonging, power, and fun through numerous church-related activities (e.g., Mass, couples outings, children events, and private Catholic school for Sam). During termination sessions, Keith reported, "I just don't seem to get upset any more with the little things—you know, the cluttered kitchen counters tops." Sue stated, "It is important to go out once or twice a month, just for Keith and me to keep our romance alive."

DISCUSSION

The case example illustrates the importance of maintaining a safe and respectful environment, as well as using particular techniques to keep the couple on task. Initially, the counselor surveyed the strengths of each partner's basic needs in order to gain some contextual information. This information allowed the counselor to clarify why each partner's preferences and positions were important. This information also underscored the similarity in goals and values that the couple shared—an important basis for resolving their presenting concern.

As noted from the case example, the counselor used other techniques and resources that were not necessarily based in RT. One of the many strengths of using RT is that it is flexible enough to incorporate and apply meaning using RT language such as WDEP, total behavior, and quality world to various techniques and assessments. Finally, the RT therapist incorporated teaching to educate the couple about the relationship between their total behavior and quality world pictures of what they both wanted out of the marriage. The psychoeducational approach of teaching the clients choice theory not only helped them address the presenting issue but will serve to help them work through any other issue that they face upon termination. This is what choice theory is all about: teaching it, sharing it, and living it. The hope is that this couple will have learned how to improve their marriage and their lives and that they will share this information with others.

REFERENCES

Bryant, C. M., Conger, R. D., & Meehan, J. M. (2001). The influence of in-laws on change in marital success. *Journal of Marriage and Family, 63,* 614–626.
Duba, J. D. (2008). The practice of marriage and family counseling and Catholicism. In J. D. Duba Onedeara (Ed.), *The role of religion and marriage and family counseling.* New York, NY: Routledge.

Duba, J. D. (2009). The basic needs genogram: A tool to help inter-religious couples negotiate. *International Journal of Reality Therapy, 29*(1), 13–17.

Duba, J. D., Graham, M. A., Britzman, M., & Minatrea, N. (2009). Introducing the "basic needs genogram" in reality therapy-based marriage and family counseling. *Journal of Reality Therapy, 28,* 69–77.

Duba, J. D., & Nims, D. (In press). Counseling Muslim couples from a Bowen family systems perspective. In D. Carter. (Ed.), *The process of Muslim family therapy: Overcoming social anxiety and acceptance.*

Fincham, F. D., & Beach, S. R. H. (2002). Forgiveness in marriage: Implications for psychological aggression and constructive communication. *Personal Relationships, 9,* 239–251.

Frame, M. W. (2004). The challenges of intercultural marriage: Strategies for pastoral care. *Pastoral Psychology, 52*(3), 219–232.

Gardyn, R. (2002). Breaking the rules of engagement. *American Demographics, 24*(7), 35–37.

Glasser, W. (1965). *Reality therapy: A new approach to psychiatry.* New York, NY: Harper & Row.

Glasser, W. (1984). *Control theory: A new explanation of how we control our lives.* New York, NY: Harper & Row.

Glasser, W. (1995). *Staying together: A control theory guide to a lasting marriage.* New York, NY: HarperCollins.

Glasser, W. (1998). *Choice theory: A new psychology of personal freedom.* New York, NY: Harper & Row.

Glasser, W. (2000a). *Counseling with choice theory: The new reality therapy.* New York, NY: HarperCollins.

Glasser, W. (2000b). *Reality therapy in action.* New York, NY: HarperCollins.

Glasser, W. (2003). *Warning: Psychiatry can be hazardous to your mental health.* New York, NY: HarperCollins.

Glasser, W., & Glasser, C. (2000). *Getting together and staying together.* New York, NY: HarperCollins.

Hirsh, S., & Kummerow, J. (1989). *Life types.* New York, NY: Warner Books.

Keirsey, D., & Bates, M. (1978). *Please understand me: Character & temperament types* (5th ed.). Del Mar, CA: Prometheus Nemesis.

Kosmin, B. A., Mayer, E., & Keysar, A. (2001). *American religious identification survey.* New York, NY: Graduate Center of the City University of New York.

Kroeger, O., & Thuesen, J. (1988). *Type talk: The 16 personality types that determine how we live, love and work.* New York, NY: Dell.

Lara, T., & Duba Onedera, J. D. (2008). Inter-religion marriages. In J. D. Duba Onedera (Ed.), *The role of religion and marriage and family counseling.* New York, NY: Routledge.

Lehrer, E. L., & Chiswick, C. U. (1993). Religion as a determinant of marital stability. *Demography, 30,* 385–403.

Minatrea, N., & O'Phelan, M. (2000). Myers-Briggs and reality therapy: Using Myers-Briggs typology in the reality therapy process. *International Journal of Reality Therapy, 19*(2), 15–20.

Myers, I., & McCaulley, I. (1990). *Manual: A guide to the development and use of the Myers-Briggs type indicator.* Palo Alto, CA: Consulting Psychologists Press, Inc.

Oklander, V. (1988). *Windows to our children: A Gestalt therapy approach to children and adolescents.* Guoldsboro, MA: Gestalt Journal Press.

Oklander, V. (2006). *Hidden treasure: A map to the child's inner self.* London, England: Karnac Books.

Tieger, P., & Barron-Tieger, B. (1992). *Do what you are: Discover the perfect career for you through the secrets of personality types.* Boston, MA: Little, Brown and Co.

Weigel, D. J., & Ballard-Reisch, D. S. (1999). The influence of marital duration on the use of relationship maintenance behaviors. *Communication Reports, 12*(2), 59–70.

Wubbolding, R. (1986). *Using reality therapy.* New York, NY: Harper & Row.

Wubbolding, R. (1991). *Understanding reality therapy: A metaphorical approach.* New York, NY: HarperCollins.

Wubbolding, R. (2000). *Reality therapy for the 21st century.* New York, NY: Routledge.

Wubbolding, R. (2011). *Reality therapy: Theories of psychotherapy series.* Washington, DC: American Psychological Association.

Zink, D. W. (2008). The practice of marriage and family counseling and conservative Christianity. In J. D. Duba Onedeara (Ed.), *The role of religion and marriage and family counseling* (pp. 55–72). New York, NY: Routledge.

11

THE STRUGGLE FOR IDENTITY AS A GAY COUPLE

Vanessa L. White and Patricia A. Robey

INTRODUCTION

Kevin and Jason are a gay couple who have been in a relationship with one another for 2 years. Six months prior to counseling, they moved in together and are now living in Kevin's apartment. Kevin, who is 30, came out to his parents and siblings when he was a teenager. His younger brother and sister were very supportive and loving about their brother being gay. Kevin's parents initially had a difficult time accepting the fact that he was gay, but are now able to accept and appreciate Kevin as a gay man who is in a committed relationship.

Jason is a 27-year-old only child. He has not disclosed to his parents that he is currently living with someone and, although they live close by, they have never visited the apartment that he and Kevin share. Jason believes that telling his parents "will only create headaches for me and my life; besides, it is no one else's business whom I am with or what I do." Jason's family and Kevin are both important to him, and this is creating deep conflict for him.

Kevin loves Jason deeply and is anxious and excited for the opportunity to meet his family. He is impatient and frustrated at the fact that Jason's family still do not even know that they are living together. Kevin is also growing more resentful about Jason's insistence that their relationship, and their love, must be "hidden" from Jason's loved ones.

Conflicts have arisen frequently as a result of Jason's reluctance to disclose his relationship with Kevin and the fact that he is gay to his family, coworkers, or friends. There are times when Jason is not even sure that he "likes" the fact that he is gay. Kevin has gradually been telling people in their lives about his relationship and has come out at work as well.

Recently, Kevin has threatened Jason, saying that if Jason is not truthful with his family, at least about the two of them living together, then he wants to break up. Jason states that he feels pressured to open up about something that he is afraid that his family will have a negative response to.

Kevin and Jason have come into therapy as a couple because they both see their relationship as being part of their *quality world* (Glasser, 1998); that is, the relationship is valued and brings a feeling of pleasure and satisfaction to them. However, things that are important to them each as individuals are currently in conflict with one another. They both would like to resolve those issues of conflict so that they can be happy as individuals and as a couple.

CHALLENGES UNIQUE TO LESBIAN AND GAY COUPLES

It is difficult to estimate accurately how many persons in the United States have a gay or lesbian sexual orientation. According to the U.S. Census Bureau (2003), approximately 1% of persons identify themselves as gay. Other statistics indicated larger percentages. Kinsey, Pomeroy, and Martin (1948) reported that 4% of men were engaged in same-sex behavior for their entire life. Kinsey, Pomeroy, Martin, and Gebhard (1953) reported that up to 3% of women identified as lesbian, and Gonsiorek and Weinrich (1991) estimated that between 4% and 17% of the population identified themselves as gay or lesbian (as cited in Hill, 2007). Regardless of exact percentage, because the majority of the population in the United States is presumed to be heterosexual, gay persons are, in number, a minority.

Similarly to other minorities, gay and lesbian individuals may experience stereotyping, social stigma, and negative reactions to their orientation by members of the social majority. However, gays and lesbians face different challenges because they were not raised and do not live in communities in which most others share their minority status. Ethnic minority is more readily identifiable than sexual orientation. The process of coming out is linked to sexuality, which is a difficult topic for many people to discuss (Israel & Selvidge, 2003).

Gay or lesbian couples seeking therapy are likely to be seeking support and assistance for what any heterosexual couple would seek therapy for, including financial issues, sexual issues, and family issues. In addition, there may be stressors that are unique to gay and lesbian persons, such as coming out to family, the degree to which each person is out to others, experiencing prejudice, and creating an authentic sense of self (Geier, 2007; Kurdek, 2004, 2005). Specific couples issues that may come up when working with a gay or lesbian couple are those in which one partner is out of the closet, but the other is not; the family of one, or both, does not know about the relationship; for lesbian couples, a tendency to lose individual identity of self; and frustration about the lack of equal legal benefits such as marriage, health benefits, and adoption rights (Group for the Advancement of Psychiatry, 2007).

If therapists are to address the specific needs, wants, behaviors, and perceptions of gay and lesbian clients adequately, they must be aware of their own attitudes and biases and be able to establish a safe and nonjudgmental environment. This friendly atmosphere provides for the effective use of therapy. Therapists should also be willing to educate themselves by reading about gay history and current trends; attending events, workshops, and training on topics related to lesbian and gay identity; and seeking out knowledge from gay and lesbian persons directly. Therapists must be aware of their own assumptions, learn to understand how their clients see their worlds, and construct case conceptualizations that are culturally relevant. They should make an effort to gain knowledge of various aspects of identity development and the coming out process, to find resources that will be helpful for clients, and to act as advocates through research and outreach (Israel & Selvidge, 2003).

CHOICE THEORY

Choice theory explains human behavior and motivation (Glasser, 1998; Wubbolding, 2000, 2010). According to choice theory, we are born with *basic needs* for love and belonging, power, freedom, fun, and survival. We meet these needs through very specific people and things that we hold to be important. The desire for these people and things is stored in our memory in what Glasser refers to metaphorically as our *quality world.* We are continually comparing what we want (our quality world picture of our ideal situation) to what we believe we actually have (our perception of our reality). This comparison drives us to behave in order to maintain what we are currently getting, if it is what we want and need, or to make changes in order to meet our needs more successfully. Therefore, our *total behavior* of acting, thinking, feeling, and physiology is purposeful and our best attempt to get the things that we want in our quality world, which satisfies one or more of the basic needs.

THE CASE OF KEVIN AND JASON

Choice theory teaches that we seek congruence between the pictures of what we want in our lives, our "quality" world, and what we experience in our "real" world. When we have this congruence, our lives feel more in balance because the two worlds are more in balance (Wubbolding, 2000). True happiness occurs when people are enjoying their lives and getting along well with the persons in their lives (Glasser, 2003). Being secure in one's sexual orientation can help in living a happier life.

Identity Development

Identity development models are based on stages of development that occur as a person identifies himself or herself as gay or lesbian and then, his or her process of coming to terms with that sexual orientation. Cass (1979, 1984) described six stages of homosexual formation. These stages range from identity confusion, or feeling different from one's peers, to identity synthesis, when a person's sexual orientation becomes part of a person's overall identity (as cited in Bilodeau & Renn, 2005). Carrion and Lock's 1997 model includes eight stages, beginning with an internal discovery of sexual orientation, which includes feelings of bewilderment, shame, feeling different, fearing rejection and abandonment, and denial, through the final stage of integrating identity between self and society (as cited in Mosher, 2001).

Choice Theory and Identity Development

Kevin and Jason both went through the process of identity development; however, their experiences were very different. As Kevin grew up, he began to realize that he had feelings in relation to other young men that were confusing to him. At first, this created a sense of incongruence for Kevin. In his quality world, which holds the most vivid pictures that create positive feelings for him, Kevin had pictures of his family. His family included his mother and father and a younger brother and sister. As he moved through adolescence, Kevin began to feel deeper feelings toward persons of the same gender, and he fell in love for the first time. When Kevin told his mother about his feelings, she struggled with coming to understand the situation and felt that it was wrong. When Kevin compared his quality world picture of his mother's approval with the disapproval she expressed that was related to his feeling of love for the young man, he felt out of balance and had an urge to behave in order to make his world more congruent. He wanted his mother's approval and he wanted the relationship with the young man.

Imagine a scale with two sides that are equal. When Kevin put his family and their approval on one side of the scale and his feelings for the young man on the other side, there would be a tipping in one direction or the other, depending on which he was focusing on as more important in that moment. It seemed that that scale could not come into balance, that these two important pictures could not be congruent with one another. Kevin's parents, through educating themselves and seeking support, were able to become more open, loving, and accepting toward him and his sexual orientation, as well as openly accept his partner, Jason, as part of Kevin's life. Kevin's current relationship with his family is positive and loving, and they are an important quality world picture for him. His picture of himself as a gay man is in his quality world; it is need satisfying and he feels happy with that aspect of his identity. In addition, Jason is in his quality world as his partner and as one of the ways that Kevin meets his need for love and belonging.

Jason's experience was quite different. Jason is an only child and has never told his parents that he is gay. Jason grew up with his parents being in his quality world. When he began having strong feelings for persons of the same gender, he experienced incongruence between his perceptions and his feelings about other young men. The information that Jason learned in his life was that being gay was wrong. This information was incongruent with what Jason was feeling, so he chose behaviors to attempt to change or hide those feelings. Some of the behaviors included dating young women, bringing them home to meet his family, and denying his feelings for other young men. When people have incongruence between a picture in their quality world and the real world in which they live, it is often referred to as their scales being out of balance. For Jason, his scales were out of balance when he compared the quality world picture of his current relationship with Kevin, which is important to him, to the world in which he lives—a world that dictates that he should keep the relationship hidden, even from his family. Because his family is also in his quality world, this is creating a deep conflict for Jason.

The Coming-Out Process

Coming out is a process unique to gays and lesbians. It has been described as the end process of restructuring identity, in which the individual acknowledges that he or she has resolved internal conflicts related to sexual identity and is ready to self-disclose sexual identity (Mosher, 2001). It is important for a therapist to have a keen understanding of coming out as a process, rather than as an event. Coming out is typically discussed as having two main components: coming out to self and coming out to others.

Coming Out to Self The process of coming out to self can be lifelong for some persons. Factors that can impact this process include internalized homophobia, which is defined as an intense, irrational fear, hatred, or loathing of homosexual persons (McWhirter, 1994). The presence of homophobia can influence how much a person has come out to himself or herself or others. Homophobia can be present in the gay person toward himself or herself or others who are gay or lesbian. Individuals who are part of a gay person's support system, workplace, and/or community can have homophobic attitudes toward that person as well. It can exist among therapists or others in the helping professions. It can also show itself in the avoidance of gay and lesbian persons or even in mistreatment and violence toward them.

Coming Out to Others When a person is coming out as gay or lesbian to others, there are several factors to consider. As part of one's identity, sexual orientation is somewhat mutable; therefore, decisions have to be made on a daily basis about in whom to confide about one's sexual identity. Unless a person is at ease with coming out to others (i.e., both gay and heterosexual persons in his or her life), that person possesses a false sense of self. This false sense of self is a result of not being authentic about who they truly are. An individual may feel that his or her only true value in the eyes of others is in what those others want him or her to be, rather than who he or she really is (Martin, 1991).

Choice Theory and Coming Out

Our quality world pictures are socially constructed within the influence of family values, culture, religion, and law, and they represent the ways that we meet our basic needs for belonging, power, freedom, fun, and survival. We may hold onto our pictures in our quality worlds, even if we are no longer able to satisfy them. People who are coming to terms with a gay sexual orientation may hold onto the quality world picture of heterosexual relationships, even if it is no longer totally need satisfying to them.

From a choice theory perspective, Kevin and Jason appear to have a high need for love and belonging. However, Jason has a low need for freedom and high need for survival. Jason believes that telling others about his being gay is risky behavior. Jason also believes that to tell his family that he is gay will move him further away from them emotionally, due to the danger of their possibly rejecting him.

Jason states clearly that his parents and his relationship with them are in his quality world. In his real world experience, there are many people with negative attitudes about persons who are gay or lesbian. Because Jason values the input of people who are important to him, he has adopted those attitudes and has come to put a negative value on his sexual orientation (internalized homophobia). He is reluctant to come out to the people in his life whose good opinions and love he craves because he fears he will lose their acceptance.

A gay or lesbian person can adopt internalized homophobic attitudes more easily if the person has been part of support systems and societal systems that possess the idea that heterosexuality is right and homosexuality is wrong. As a result, people who grow up to have a gay or lesbian sexual orientation can have certain negative feelings about themselves because they feel different from other people around them (Herek, Cogan, Gillis, & Glunt, 1997). This difference comes from the incongruence between what is seen as the "ideal" quality world picture of heterosexuality and their own feelings of homosexuality, which is less than "ideal." People may choose behaviors such as depressing to cope with the incongruence in their perceptions. Higher levels of internalized homophobia are correlated with lower self-esteem, higher rates of depression, and decreased likelihood of coming out to others (Herek et al., 1997).

Choice theory explains that when fear and negative thoughts are directed inward, people are not likely to feel in effective control of their lives. The incongruence between what is perceived versus the quality world picture creates such painful feelings that individuals do what they can to gain control of their lives. Even when our behaviors appear self-destructive, it is always our best attempt to get what we want, which satisfies our needs. In this case, turning hatred toward self is one behavior gay persons may choose in order to manage the incongruence that they are experiencing. As a result, some gay and lesbian persons deny their sexual orientation to themselves or to others (Silver, 1997).

At times, this may result in denying gay feelings and maintaining heterosexual relationships, marriages, and families, in order not to have to face up to those feelings. This choice can be effective in that the person is now matching the picture of what he has accepted as being the social ideal and the way life *should* look. In this way, the behavior is need satisfying, which is why he is choosing it. Passing as heterosexual may also be a technique for self-preservation because violence and hostility are often directed toward gay and lesbian individuals (Mosher, 2001). Therefore, the choice to "pass" meets the need for survival and safety. On the other hand, this choice is ineffective, in that other needs, such as the need to have a relationship with an individual of the same gender, are not being met.

REALITY THERAPY

Reality therapy is grounded in choice theory and is the process used to help clients find more effective ways to get control of their lives (Glasser, 1965; Wubbolding, 2000, 2010). Although people have a variety of different problems, Glasser (1998) stated that the real long-term problem is always rooted in an unsatisfying relationship. The difficulty we have in our relationships is usually because we are trying to control the important people in our lives. We do this through the use of what Glasser and Glasser (2000) call the *deadly habits:* criticizing, blaming, complaining, nagging, threatening, punishing, and bribing or rewarding to control.

Reality therapy works in the present and focuses on helping clients define what they want (quality world pictures that meet basic needs), rather than focusing on what they do not want (the problem). The therapist helps clients identify the behaviors they have been choosing to get what they want and need and then to evaluate the effectiveness of this behavior. Finally, the therapist helps clients create a plan for more effective behavior. As part of the counseling process, clients are taught to substitute *caring habits* (listening, supporting, encouraging, trusting, respecting, accepting, and negotiating differences) for the deadly habits in their relationship. Clients are also taught the concepts of choice theory so that they can use these ideas to improve their relationships and to take more effective control of their lives (Glasser, 1998; Wubbolding, 2000, 2010).

REALITY THERAPY IN THE CASE OF KEVIN AND JASON

Kevin wants to be open and honest about his love for Jason, and he is feeling unbalanced because in his quality world, there is a picture of his relationship with Jason. When Kevin observes what is happening in his real world, his perception is that Jason does not want to acknowledge their relationship, which Kevin believes indicates that Jason does not value it. Both of the men have been using external control and deadly relationship habits, including nagging, threatening, and criticizing one another, in order to try to get the other to change. This has not worked, and their relationship has suffered as a result. They both are coming in for therapy because they are ready to make changes that will enhance the relationship and enable them to stay together as a couple.

Session I

For the first session, the therapist uses *structured reality therapy* (Glasser, 2000), a variation of the reality therapy process that can be used with couples or groups. This process is highly structured, with allowances for both parties to be able to contribute to the task of discovering what is not working in the relationship and to make plans for change. Through the use of the following five steps, Jason and Kevin can begin to discover what it is they are willing to do to save their relationship.

Kevin and Jason come in for their first session. After beginning to build a relationship with the couple, the therapist asks the first question in the structured reality therapy format.

Question I   Therapist: *Are you here because you want help for your current relationship?*

When both Kevin and Jason answer yes, the session continued. This was the first step in moving forward in the use of structured reality therapy (Glasser, 2000). If one of them had said no, then the session would have ended, and they would have been asked if they would be willing to come back and try again at a future time.

Question 2   Therapist: *Whose behavior can you control?*

With some reluctance, Kevin and Jason acknowledge that they can only control themselves. The therapist then teaches Kevin and Jason about external control psychology, how it works, and what relationship-breaking habits humans use to exert control over others. Kevin and Jason listen intently to the definition of external control psychology, which maintains the belief that a person can control another person's behavior through behaviors such as blaming, criticizing, and threatening. Choice theory explains that this belief is incorrect, that we can only control our own behavior, and that the use of behavior to try to control others only damages relationships and does not influence long-lasting change (Glasser, 2003). Kevin and Jason acknowledge that they have used coercive behaviors in an attempt to control one another. This has only brought about temporary change, and, as a result, their relationship is strained and unhappy.

The therapist instructs Kevin and Jason that for the remainder of the session, they are not to use any external control behaviors, such as nagging, criticizing, or punishing. In order to create effective change in their relationship, Jason and Kevin need to accept this and work within the context of having control only over their own individual behaviors.

Question 3   Therapist: *Would each of you tell me what you believe is wrong with the relationship right now?*

In order to understand the perceptions of Kevin and Jason regarding their relationship, each is asked to share, one at a time, what he believes is wrong with their relationship. Kevin states that the problem is that Jason has not come out to his family or to his coworkers or other friends. Kevin also states that Jason does not seem to want to be gay at times, as if it is a hassle that he does not want to deal with. Kevin states that Jason seems resentful and is acting negatively toward him over these issues; as a result, they are either fighting or not speaking most of the time.

Jason responds. He states that he feels intense pressure from Kevin to be out in the open about being gay and about their relationship, and he is resentful about that. He believes that his family will respond negatively if he tells them that he is gay, and he fears they will disown him. Jason also states that the fighting and not speaking to one another has created a big strain on the relationship and that he is afraid that Kevin will leave him.

Question 4   Therapist: *In your opinion, what is good about your relationship right now?*

The therapist attempts to shift the couple from the problem focus to the areas in which the relationship is working. This creates a change in emotional energy that provides impetus and strength for thinking about solutions rather than problems. Both men are asked to share what they see as the positive aspects of their relationship. Kevin states that he appreciates Jason's sense of humor; that they laugh together a great deal; have the same interests, such as hiking and sports; and enjoy traveling. Jason states that he appreciates Kevin being open about his feelings, that he wants to let Jason know events in his life, and that Kevin's family has been open and loving toward him.

Question 5   Therapist: *Tell me one thing each of you could do all this coming week that you think would make your relationship better.*

Jason and Kevin are then asked what each of them is willing to do in the next week to help save their relationship. Now that the couple is feeling more positive toward one another, they are more likely to be creative in generating effective behavior changes. The focus of therapy is to help clients eliminate their ineffective behaviors and replace them with behaviors that are likely to bring the couple closer together. As the couple experiences more successful interactions, this behavior is likely to increase and to have a positive influence on the relationship.

Jason says he understands the need to be more open about his love for Kevin and agrees to tell one of his friends that he is gay and in a relationship before the next session. Although Jason does not feel that he is ready to tell his family, he does feel that he wants others to know how happy he is. Kevin also agreed to compromise and to do a task that would help the relationship. He agrees that he will spend the week without threatening Jason with leaving, and he will not pressure Jason about his family and coming out to them. Kevin also states that he will go along with Jason when he comes out to his friend, if Jason would like him there. Both Kevin and Jason agree to follow through on their plans and to return for their next session with an update.

Session II

Kevin and Jason return 2 weeks after their initial session to report on their progress. Jason is proud to report that he did go forward and tell a friend of his that he is gay, and that his friend was very supportive. Jason felt very positive about the experience. However, since coming out to his friend, Jason has not made any other attempts or hints that he plans to come out to anyone else in his life in the near future. Although Kevin was initially quite supportive of and encouraged by Jason's coming out to his friend, he is increasingly frustrated regarding the fact that Jason does not continue to let others in his life know that he is gay and to acknowledge their couple relationship to others. He began nagging Jason again and once more threatening to leave him if this situation did not improve. Jason, however, acknowledged that he did want to save the relationship and that he fully accepts his gay identity and wants to resolve his fears about coming out.

The Solving Circle　　The second session begins by explaining the solving circle (Glasser, 1998). Whether literally a circle in the room or simply a symbol of the commitment, the solving circle is the place in which both individuals in the couple agree to work on their relationship with one another. The basic premise is that an imaginary circle can be created in the therapy room or in the home. When the couple wants to address issues in the relationship, both individuals enter the circle, in which are three entities: each individual and the relationship. The understanding is that when the couple is facing a problem, both individuals agree to enter the circle to discuss it. They are admitting that the relationship needs are more important than their individual needs. There is no room for conflict or coercion in the circle, and both individuals have to be willing to state clearly what they are willing to do to improve the relationship.

The main goal here is compromise. It is important for Jason and Kevin to remember that even though they are frustrated about the state of their relationship, they both entered the circle to put the relationship ahead of their individual demands and needs. If either one of them decides that this is not true for him, they will leave the circle and possibly return later. Kevin and Jason agree that the relationship is the primary focus and agree to stay in the circle and come to a compromise.

Jason speaks first. He states that he has gained understanding, by coming out to a friend, of how important it is to feel confident in his gay identity and states that he believes that he would benefit by receiving therapy on his own to help with that process of self-acceptance. He states clearly that his relationship with Kevin is in his quality world, and he wants to maintain it as healthy and happy as possible.

Kevin is quite pleased with this compromise that Jason has proposed because his belief is that it not only will help their relationship, but will also help Jason individually. Kevin acknowledges his own impatience with Jason throughout this adjustment, and he agrees that he will no longer nag Jason to tell others that he is gay and will, instead, encourage Jason to come as often as possible with Kevin to visit his family and their friends who know that they are a couple.

The therapist gives Jason a referral for an individual therapist, and the men agree to return in 1 month to report on their progress. The therapist recommends that Jason and Kevin remember that they are actively working within the context of the solving circle for their relationship, even outside the confines of the office, which will give the most hope to the growth and development of their relationship. They are also given a list of the seven deadly and caring relationship habits so that they can evaluate the effectiveness of the behaviors they are using with one another.

CONCLUSION

Choice theory and reality therapy are concepts and tools that can be applied to any couple relationship. The needs of gay and lesbian couples are specialized in some specific ways; however, if the needs of both persons can be identified, and both are willing to work on the relationship as their commitment, the relationship can be stronger and more need fulfilling for both individuals. Through the use of choice theory, reality therapy, structured reality therapy, and the solving circle, issues that have created conflict can be resolved more readily and replaced with the use of the caring habits.

REFERENCES

Bilodeau, B. L. & Renn, K. A. (2005). Analysis of LCBT identity development models and implications for practice. *New Directions for Student Services, 111,* 25–39.

Cass, V. C. (1979). Homosexual identity formation: A theoretical model. *Journal of Homosexuality, 4,* 219–235.

Cass, V. C. (1984). Homosexualtiy identity formation: Testing a theoretical model. *Journal of Sex Research, 20,* 143-167.

Geier, P. (2007). *Couples therapy for lesbians and gay men: The basics.* Retrieved from http://www.goodtherapy.org/blog/couples-therapy-for-lesbians-and-gay-men-the-basics/

Glasser, W. (1965). *Reality therapy: A new approach to psychiatry.* New York, NY: Harper & Row.

Glasser, W. (1998). Choice *theory: A new psychology of personal freedom.* New York, NY: HarperCollins.

Glasser, W. (2000). Reality *therapy in action.* New York, NY: HarperCollins.

Glasser, W. (2003). *Warning: Psychiatry can be hazardous to your mental health.* New York, NY: HarperCollins.

Glasser, W., & Glasser, C. (2000). *Getting together and staying together: Solving the mystery of marriage.* New York, NY: Harper Collins.

Gonsiorek, J. C., & Weinrich, J. D. (1991). The definition and scope of sexual orientation. In J. C. Gonsiorek & J. D. Weinrich (Eds.) *Homosexuality: Research implications for public policy* (pp. 1-12). Newbury Park, CA: Sage.

Group for the Advancement of Psychiatry (2007). *Psychological development and life cycle.* Retrieved from http://www.aglp.org/gap/3_development/

Herek, G. M., Cogan, J. C., Gillis, J. R., & Glunt, E. K. (1997). Correlates of internalized homophobia in a community sample of lesbians and gay men. *Journal of Gay and Lesbian Medical Association, 2,* 17–25.

Hill, C. A. (2007). *Human sexuality: Personality and social psychological perspectives.* Thousand Oaks, CA: Sage Publications Inc.

Israel, T., & Selvidge, M. M. D. (2003). Contributions of multicultural counseling to counselor competence with lesbian, gay, and bisexual clients. *Journal of Multicultural Counseling and Development, 31,* 84–98.

Kinsey, A. C. , Pomeroy, W. B., & Martin, C. E. (1948). *Sexual behavior in the human male.* Philadelphia, PA: W. B. Saunders.

Kinsey, A. C. , Pomeroy, W. B., Martin, C. E., & Gebhard, P. H. (1953). *Sexual behavior in the human female.* Philadelphia, PA: W. B. Saunders.

Kurdek, L. A. (2004). Are gay and lesbian cohabiting couples really different from heterosexual married couples? *Journal of Marriage and Family, 66,* 880–900.

Kurdek, L. A. (2005). What do we know about gay and lesbian couples? *Current Directions in Psychological Science, 14*(5), 251–254.

Martin, H. P. (1991). The coming-out process for homosexuals. *Hospital and Community Psychiatry, 42*(2), 158–162.

McWhirter, E. H. (1994). *Counseling for empowerment.* Alexandria, VA: American Counseling Association.

Mosher, C. M. (2001). The social implications of sexual identity formation and the coming-out process: A review of the theoretical and empirical literature. *Family Journal, 9*(2), 164–173.

Silver, D. (1997). *The new civil war: The lesbian and gay struggle for civil rights.* New York, NY: Grolier Publishing.

U.S. Census Bureau 2000. (2003). Married-couple and unmarried partner households: 2000.

Wubbolding, R. (2000). *Reality therapy for the 21st century.* New York, NY: Routledge.

Wubbolding, R. (2011). *Reality therapy: Theories of psychotherapy series.* Washington, DC: American Psychological Association.

12

THE CELEBRITY CHALLENGE
Counseling High-Profile Clients

Brandi Roth

INTRODUCTION

Being a therapist to prominent individuals is a privilege, but not all glamour. This counseling practice has its own set of challenges and complexities, especially in couples counseling, when one partner is a celebrity or in a high-profile position and is accustomed to being idolized, catered to, and in professional control, while the other partner is less prominent. Helping clients learn to recognize and understand the purpose behind controlling and dominating behavior is an important first step in the counseling journey.

Choice theory (Glasser, 1998) and reality therapy (Glasser, 1965) are especially valuable additions to the couples counseling process because they work hand in hand to provide a comprehensive approach to the therapeutic experience. Choice theory explains how and why we are internally motivated to generate behaviors and is the basis for reality therapy—the procedures that lead to change. Using reality therapy, clients are asked questions to examine their needs and their quality world pictures, or to identify what they want to change.

Using the choice theory concept of total behavior (Glasser, 1998), clients are asked what they are doing to get what they say they want. They examine whether their behavioral choices are helping or hurting their chances of having a good relationship. When this self-evaluation process is effective, clients are encouraged to be creative in making new plans to solve their relationship problems. Creativity, as explained in *Choice Theory* (Glasser, 1998), is a significant component of the counseling process. Reality therapy questions lead clients to make creative plans to change the direction of their lives. Reality therapy helps individuals evaluate choices and make plans that lead to happier relationships.

THE WHOLE WORLD IS WATCHING: VISIBILITY VERSUS PRIVACY

A by-product of fame is hypervisibility and constant recognition. Feeling that they know them from media-based information, the public often acts overly familiar with celebrities when encountering them in real life. Strangers approach and speak to famous people as casually as they would to an acquaintance, often acting quite uninhibited about asking personal questions, offering comments, or invading personal space. Consequently, celebrity couples must contend with the tension between notoriety and the desire for privacy.

Adapting to a life of prominence can be particularly difficult for spouses and children, who often feel overexposed and sometimes invisible, as well as undervalued. The activities of daily living are often restricted because of the connection to the famous family member. This can lead to resentment and relationship tension. In counseling with reality therapy, the therapist guides prominent couples to evaluate their thinking and actions. The goal of this self-assessment is to find ways to develop and maintain closer interpersonal relationships, expand communication, and plan social events that every family member can enjoy.

TREATING CELEBRITY AND HIGH-PROFILE CLIENTS

Establishing an initial therapeutic relationship with a couple includes gathering information about the state of the relationship. If they are considering separation or divorce, they sometimes want a collaborative resolution to their disputes, although they might have a hidden agenda to strengthen their litigation. Pressure to settle can be greater for celebrities than noncelebrities. They have limited opportunity to litigate with privacy or anonymity. Adverse publicity or media scrutiny can harm a celebrity more than a noncelebrity. Clients in counseling who are in litigation with each other often expect the therapist to determine both the stated and unstated agendas and to distinguish between unrecognized and intentionally hidden goals. Prominent individuals often talk to close friends and business consultants as advisors. Telling clients what they want to hear can make settlement less likely and the role of the mediator more difficult. Therapy and mediation can blend, but only if the clients and the therapist formally agree to a clearly defined dual role.

It is imperative that the therapist/mediator has a high degree of factual accuracy before making decisions or recommendations that appear to support the position of one party and appear to reject the position of the other. Occasionally, couples ask the therapist to assume the role of dispute mediator, suggesting decisions for the clients. This can be a challenging and complex position, especially in high-conflict relationships where obtaining sufficient facts and neutral information from which to draw proper conclusions can be difficult. The therapist should work only within her or his area of expertise and level of specialized training. It is important to recognize when to collaborate with other professionals, when to refer to other colleagues, and/or when to seek medical or legal advice.

SELF-PERCEPTION OF CELEBRITY AND HIGH-PROFILE CLIENTS

In celebrity or high-profile relationships, there is often a dichotomy between how the famous see themselves as reflected through the eyes of their adoring office or public and how they are perceived through their spouses' or partners' eyes. Celebrities can live in a smaller universe with a big power differential. Their prominence can mean they are highly recognizable, highly valued, or in a position where people submit and acquiesce to them. For an unknown or lesser known celebrity in a relationship with a celebrity, there are often pressures that result from having lesser visibility and status in the work place or society. They may have the perception of being a prominent "add-on" to the celebrity. In a relationship where both partners are public figures, one might outshine the other. This can create issues of jealousy, a loss of control and power, or a feeling of being minimized. Over time, the pendulum can swing both ways.

Faults and foibles can descend into bickering and blaming. Nevertheless, most prominent couples learn to make adjustments to the balance of power within the relationship. When celebrities, who are accustomed to exerting their will on others, ask for help, the usual strategies for therapeutic boundaries, confidentiality, and frameworks for change must be adjusted to account for celebrity issues.

Celebrity clients in couples therapy frequently concentrate on and proceed under one of three main scenarios:

- Addressing the how-to of maintaining a satisfying couple relationship while enjoying individual success and fame
- Addressing topics such as satisfying each person's needs and reconciling conflicts between perceptions and individual behavior
- Resolving entanglements if there are secretive behaviors in the relationship

Topics relevant to prominent couples frequently include:

- How each client's history impacts present needs and wants
- Each partner feeling heard and understood when telling his or her autobiographical narrative
- Understanding star power as it relates to status, talent, creativity, and the impact on a partner
- The challenge of transitioning between work and personal or family relationships
- Strategies to build connected relationships with both independence and interdependence
- Understanding the neuroscience behind feelings, physiology, thinking, and action (components of total behavior) in order to gain understanding of behaving
- Reassessing collaboration that is effective or ineffective (helpful or hurtful)
- Understanding choices for new behaviors
- Road maps for making a new plan
- Understanding the needs for survival, freedom, fun, power, and love and belonging

The purpose of using reality therapy and teaching clients choice theory is to make a difference in couples' lives by facilitating change, teaching effective strategies, and finding new and creative action plans. Therapists using reality therapy have an expanded toolbox for ways to listen and focus on helping.

CASE STUDY OF JUSTINE AND MIKE

To demonstrate the use of choice theory and reality therapy in counseling couples, we will consider the hypothetical case of Justine and Mike and their family system. Any similarity to an actual couple or family is coincidental. They have been married for 13 years and have two children, Adam, age 12, and Tiffany, age 10. Like many couples, their problems are exacerbated by the responsibility for aging parents and the changing needs of their children. They report that they frequently avoid meaningful conversations with each other because they have learned over the years not to trust the direction of their conversations. They are essentially disconnected from each other and have no tools for how to resolve their differences.

Justine and Mike are not contemplating separating or divorcing and are not engaged in any litigation.

Justine is an actress and is paid extraordinary sums of money for her work. She is highly recognizable when she is out in public and is a constant target of autograph seekers and invasive paparazzi. Mike is a writer and also an entertainment figure. He is less recognizable in public and works alone or with a partner at home. He has the primary responsibility for organizing the household and the children's schedule. At work, Justine's every request is granted, her every idea is treated by her team as brilliant, and her every action is described by her staff as perfect.

At home, the story is different. She and Mike bicker frequently. Mike freely expresses his opinions to Justine, sometimes disagreeing with her ideas and her actions. Justine often interprets this as unfair criticism. Mike is frustrated by what he sees as Justine's inability to see through all the "yes" people surrounding her at work. At the beginning of the therapy, Justine and Mike initially selected coparenting differences as a topic of discussion. They had conflicting parental approaches to Adam's very strong desire for less restriction over computer and game use, bedtime, and study time. Mike supported more freedom to allow Adam to learn from his mistakes. Justine wanted to continue to set precise rules and limits for Adam. She supported strong parental oversight and control of both children.

REALITY THERAPY

To build the counseling relationship, a simple question for starting the session might be, "What do I need to know about you in order to help you today?" Justine and Mike initiated counseling thinking that their problem was about their son, Adam. The therapist asked them each to explain their concerns. After listening to their stories, the therapist opened a discussion about ways to bridge the gap between their differences regarding Adam's homework and computer time. The therapist asked the couple to share their ideal pictures of what their lives would be like if things were going the way they wanted them to be. Justine and Mike agreed that they would like to spend less time arguing and more time being together enjoying family time. The therapist helped the couple recognize both Adam's and Tiffany's efforts and time spent on schoolwork. She then helped the couple develop new ways to have the type of family evenings they desired.

As the therapist continued to work with the couple, she introduced the concept of connecting (caring) and disconnecting (deadly) habits (Glasser & Glasser, 2000). She taught the couple about the importance of language in presenting their viewpoints. The therapist gave the couple a small poster of the caring and deadly habits, which Mike posted in the living room as a reminder to the couple of how they wanted to interact with one another.

The list of caring and deadly habits also reminded Justine and Mike to listen to their children's ideas and viewpoints. Instead of dictating rules, they increased their collaboration and discussion about the times Adam would study without music, noise-making devices, or interruptions from text messages and e-mails. It was agreed that a family "study hall" time would be set aside to work on projects, with all entertainment devices turned off. Study hall would be a specific time when the whole family agreed to engage in work or study as an additional way for the parents to model effort and enjoyment of projects. Through this collaboration, Justine was also reminded that although she had the sole power to make decisions at work, home decisions were a joint responsibility. As Justine and Mike experienced the benefits and success from making collaborative agreements, they learned new ways to apply direction to other aspects of supportive guidance with their children.

Other relationship issues were identified. Justine wanted to be respected and appreciated for her hard professional work and resulting high income. Mike wanted recognition for his home and parenting efforts, many of which were not often observable. Through counseling, Mike and Justine were able to have open discussions about their individual strengths, weaknesses, and contributions from the point of view of effort, rather than money and fame. They decided to allocate time to do chores together. They included the children as part of their family time, calling it "Fun Doing Chores Together." Jobs were divided into those that were expected and essential for the family and those that were optional, creative projects. This one change led to a more satisfying connection for both partners. Their resentment of one another greatly decreased and their fun together increased.

In Justine's case, a series of life experiences as a young adult actress resulted in her description of herself as "learning to fly solo." She learned at an early age what it meant to take control and how not to depend on others for survival. In her quality world, Justine had a picture of herself as capable in social situations. By contrast, when family members became emotional, chaotic, or selfish and when hurtful behaviors emerged, Justine became deeply affected. She often got embroiled in the chaos, expecting that she could change the way they were acting. When others did not comply, her thinking and actions could become frozen in frustration. She became ineffective in communicating her needs. For his part, Mike could not understand why she broke down in the face of challenging family or social situations while, at work, she remained steadfast, organized, logical, and highly effective.

Justine learned to step back, to use a choice theory framework of total behavior (Glasser, 1998) to review her physiology, feelings, thinking, and actions, and to recognize how the behavior of other people affected her. This led her to pause, reflect, and review what she needed and wanted. She realized the only choice for controlling behavior was to control her own behavior because she did not have the ability to control others' behavior. When she felt out of control, Justine pictured herself standing on the ledge of a balcony observing and waving before leaping into the abyss below. She said this image was easy to remember as a way to stand anchored on her own psychological balcony and to wave to the chaos of others instead of jumping into their chaos. She learned to self-reflect and to evaluate her efforts and actions. This also reminded her to take time to understand the mind-sets of others. This helped Justine stay regulated, in more effective control, to keep her own mind-set, and not to be controlled by the emotions of other people.

Genogram: Incorporating History Into the Present

Choice theory teaches that we have options for how we let our history and our life experiences affect our present behaviors and our reflections on the past (Glasser, 1998). Although the focus of reality therapy is on the present, the genogram can be a useful addition to reality therapy procedures because it can be used to identify issues from the past that are alive in the present. Drawing a genogram and asking about relationships provides a revealing description of a client's family history.

A genogram is a map of the client's family that is drawn by the therapist. It is a visual extended family tree that identifies at least three generations. Immediate and extended family are named inside squares representing males and circles representing females. Connecting lines show the relationships that are close, distant, or in conflict. The drawing includes names, ages, and dates; marriages, partnerships, separations, and divorces; children, parents, siblings and significant others. Notes on the page tell the salient pieces of the family structure (McGoldrick, Gerson, & Petty, 2008). The history of relationships has an important effect on the present, and understanding the stories can be essential to charting the course of the current relationship.

Genograms provide options to record vast amounts of additional information including pertinent medical history, habits, connections, disconnections, traumas, conflicts that impacted a client's childhood, and conflicts that continue to impact the present. The genogram is referred to in subsequent therapy sessions, often while exploring history and family relationships. Additional information can be revealed, added, or revised in new ways or with added insights. A genogram can be a useful reference tool throughout the course of counseling. It can tell a narrative of the family perspective through the experiences and perceptions of the client. Clients report that the reframing of their stories by the therapist helps them move beyond their history into the present with new decisions and new strategies for more connected and happier relationships.

Ongoing Assessment

The process of creating a genogram with the therapist at the beginning of counseling often creates a collaborative connection. The therapist's ability to add insight and interpretation of the client's history based on the components of choice theory, such as needs, wants, behaviors, and perceptions, is demonstrated early. To help ensure that therapy is meeting its goals, the clients are periodically guided to reflect on reality therapy questions such as:

- What do you want/need?
  - What are your current goals for counseling?
  - What changes do you want?
  - What would you like your partner to know?
- What are you doing to get what you want? How are you acting, thinking, feeling? What is your physiological response to your situation?
  - What is your current perception of the family situation?
  - What personal change has occurred in each person?
  - In what ways do you feel understood and misunderstood?
- Is what you are doing helping or hurting your efforts to get what you want?
  - How are you genuinely supporting each other?
  - How are your individual needs being recognized?
  - What efforts have you made to recognize and respond to your own needs and your partner's needs?
  - What previously comfortable or need-satisfying behavior do you now recognize as hurtful or unhelpful? How have you changed?

- What can you do differently? What is your plan?
  - How do you plan to continue to improve?
  - What is your next objective?
  - What goals, agreements, and planning need clarification?

Opening and Closing Subsequent Sessions

Subsequent therapy sessions begin with a review of prior sessions, prior ideas, actions that have been implemented, successes and failures, and making a plan for meeting needs and wants in the current visit. By implementing self-evaluation strategies and by taking responsibility for their own behaviors, clients often benefit from a review of recent changes in lingering hurts, angers, fears, and guilt. Deeper connection occurs between couples when there is personal accountability, greater effort to express understanding of each other with appreciation, and making better choices for behaving. This sets the stage for staying in the present in the current session.

When closing each session, the therapist asks for feedback:

- What plans have been agreed upon?
- Which ideas will be implemented?
- What will be told to the children, other family members, and involved friends?
- What is the plan for the next session?

Understanding and Valuing Effort in a Relationship

The case of Justine and Mike illustrates several principles applicable to couples counseling, leadership consulting, and relationships in general. In the workplace, effective leadership requires management skills, ethics, competence, honesty, strategic planning, and collaboration. Many of these qualities overlap in personal relationships; however, the efforts needed to make a personal relationship run smoothly may not be recognized by one party and therefore may not be valued. Individuals differ in their opinions of what constitutes the predominant qualities that nurture a successful relationship. Time and effort are required to manage a household, run errands, do chores, supervise children, nurture children, cook, clean, supervise housekeepers, organize a social calendar, and manage a full social life. When done well, this leads to a comfortable and smooth-flowing lifestyle. Partners do not always recognize and value these important contributions to the relationship. Connecting these quality world pictures is one of the helping roles of the therapist.

Valuing and viewing the contributions of each person fairly can be particularly difficult when one partner is a celebrity, contributing financial largesse and social status, and the other partner lives in the background. Value can equate to perceptions of power in a relationship. Justine and Mike increased their efforts to value their partner's contributions and to show appreciation to each other, thus equalizing the power imbalance in their relationship. A successful interaction with caring behaviors increased their sense of connection.

Perceptions and Quality World Pictures

Therapists working in couples counseling face unique challenges, including helping clients understand their quality worlds, total behavior, and perceptual systems as well as how these elements impact the family as a whole. It is particularly important to clarify the strength of the perceptions of each other's situations and beliefs. Couples with children have the additional responsibility to help their children develop and learn ways to get along with siblings, parents, teachers, the extended family, and friends in a constantly changing environment.

Successful couples counseling helps people understand how the uniqueness of their thinking, feeling, and actions affects their partner. They learn how a unique set of historic life experiences influenced the way they now behave, make current decisions, and treat their partner.

Using Questions to Initiate Planning

For Justine and Mike, choice theory concepts provided the framework for the therapist to deliver reality therapy questions to expand their thinking and to explore their perceived quality world pictures (Glasser, 1998; Wubbolding, 2000, 2011). Justine and Mike were asked to think about their responses to the following questions in order to focus on their own self-evaluations and to formulate their thoughts for further discussion and resolution:

- What do you want that you are not currently getting?
- What is your ideal picture for more effective behaviors?
- What are you currently doing to get what you want?
- What areas do you have direct control over? What do you have indirect control over? What can you not control?
- What situations do you view with a new, more effective strategy?

Justine and Mike presented their responses to the questions and worked in tandem with a problem-solving framework sheet (Roth & Van Der Kar-Levinson, 2002) to summarize their stories, their perceptions, their misunderstandings, and their current dilemmas. Building on the creativity and new insights gained from the collaboration of Justine and Mike and the counseling, the couple was able to create a new plan of action.

Problem-Solving Framework

The counselor has a vital role helping clients understand each other. Providing a problem-solving framework opens a conventional exchange of perceptions and an opportunity to meet a quality world of needs and wants (Roth & Goldring, 2008). Giving clients problem-solving strategies (see Table 12.1) in a handout provides a reference explaining how the connecting process progresses from determining the situation and elaborating to action and a plan. This approach creates the freedom to share information safely and instills the confidence that problem solving will be followed by better communication. Dilemmas are presented in three steps:

1. What is the situation or information?
2. What are the behavioral symptoms that are occurring (i.e., thinking and feeling)?
3. What is the suggestion for a new choice?

This framework increases effectiveness by encouraging self-evaluation, conversation, and creativity while making a plan to change direction and behavior.

Table 12.1 Problem-Solving Chart

Step 1: Past	Step 2: Present	Step 3: Future
Dilemma	**Analysis**	**Action**
What is the information?	Total behavior symptoms:	Plan what can you do next
What happened?	What are you doing now?	Plan accountability
What are the topics of discussion?	What are you thinking now?	Make a new choice
		Change behavior
What are the situations?	How do you feel now?	Take responsibility
	What are your body sensations and reactions?	Self-evaluate
		Apologize and repair

Sources: Roth, B., & Glasser, C. (2008). *Role-play handbook: Understanding and teaching the new reality therapy counseling with choice theory through role-play.* Beverly Hills, CA: Association of Ideas Publishing; Roth, B., & Goldring, C. (2008). *Relationship counseling with choice theory strategies.* Beverly Hills, CA: Association of Ideas Publishing; Roth, B., & Van Der Kar-Levinson, F. (2002). *Secrets to school success; guiding your child through a joyous learning experience.* Beverly Hills, CA: Association of Ideas Publishing.

When couples feel safe, secure, and close, they are able to self-evaluate their individual behavior and to interact happily together by using active listening skills and paying attention to each other's facial and body language, emotions, and expressions of needs and wants. Each partner can learn to understand the other's mind-set by speaking only for herself or himself and by accepting responsibility for his or her own actions. They can feel comfortable sharing truths, knowing it is their perception of their quality world and how their needs and wants can be met.

When couples become stuck in a mind-set of feeling misunderstood, the reactions that result block problem solving. When conflict is viewed from a neutral perspective, the three steps in the problem-solving chart (Table 12.1) facilitate ways to organize and present a dilemma concisely. Couples report experiencing success at succinctly telling their story and feeling heard, often for the first time in a long while.

The Brain: Neuroscience and Total Behavior

Neuroscience discoveries continuously produce new understanding of the brain and the mind, revealing that the brain itself is an interconnected network of processes that regulate and control the body's active sensory system and transmit information that results in behavior. Within the brain, the human mind continuously organizes this network of thinking, perceptions, emotions, and bodily responses combined with both memory and creativity (Sweeney, 2009).

Choice theory describes behavior from a psychological perspective as a network of feedback loops that provide signals that influence behavior. Glasser (1998) describes total behavior (feelings, physiology, thinking, and action) as occurring simultaneously. This parallels the biological discovery that shows that the left and right hemispheres are interconnected through the corpus-callosum and brain stem (Sweeney, 2009). Glasser (1998) refers to quality world pictures that motivate individuals to organize and reorganize behavior and decisions effectively or ineffectively. These ideas match the neuroscience that describes the memory system, the sensory system and the neural networks encoding all of our behaviors and causing reactions, responses, and internally motivated choices throughout the brain (Cozolino, 2002).

Justine and Mike were shown a model of the physical brain and were taught the ways in which their emotions (such as anger or anxiousness) and sensory reactions (such as heart palpitations, headaches, or stomachaches) led to unregulated and ineffective thinking and actions (Carter, Aldridge, Page, & Parker, 2009). Paralleling this science with choice theory, total behavior serves as a powerful contribution to couples' understanding how mind-sets are affected by both early and current experiences that then lead to actions related to both satisfying and unsatisfying relationships. Reframing the manner in which information travels through the circuitry of the brain—producing new strategies and more effective choices—facilitates the brain's ability to organize information more creatively and effectively (Siegel, 2010).

Self-awareness and self-evaluation provide the cornerstones for successful choices, ways to behave, and how to change symptoms. Relationship connectedness results when there is ability to process information, communicate together helpfully, and understand each other with empathy, collaboration, awareness, and action. Using the total behavior car analogy (Glasser, 1998), Justine described driving her behavior on her back wheels. She said, "It's as if I've been looking over my shoulder toward the back seat and not paying attention to the road ahead." She went on to express a desire to use her new understanding of listening to information and to process Mike's perceptions by reviewing that information neutrally before reacting. Acting on this plan, Justine reported a positive effect on her ability to self-evaluate, to make decisions, and to respond.

Both Justine and Mike expressed their appreciation for new strategies that enabled them to share perceptions and quality world pictures with each other. Mike recognized that his response of feeling helpless to change a situation added to his own stress, as well as to Justine's misunderstanding of his actual perceptions and quality world picture of their family, his own needs, and his behavior. By understanding his emotional arousal system and his frequent tendency to be on heightened alert, but at the same time ruminative or reactionary, Mike was able to understand the manner in which his mind "tricked" his brain, leaving him without the necessary tools for interdependence. When Mike understood the circuitry of the flight, fight, or freeze emotional response, which is also known as the conscious or unconscious body alert system, he began to change his responses by describing his thinking and his understanding of a situation as it arose, rather than merely reacting. His self-regulation positively impacted his ability to build trust and collaboration and to change his actions and words. Mike and Justine moved closer to a place where each felt more appreciated by and connected with each other.

Mirror Neurons and Active Listening Skills

Relatively new neuroscience is exploring mirror neurons in the brain. Mirror neurons are thought to be related to the human ability to connect, to love, and to understand another person's experience. Mirror neurons in the brain control how individuals relate to each other by providing the dual focus of awareness of the self and others. Mirror neurons fire when people observe others. They specialize in carrying out and understanding actions and reactions. They also are instrumental in providing a reflection on other people's intentions and the social meaning of their behavior and emotions (Cozolino, 2002). With so much attention constantly flowing in their direction, celebrities can become unaccustomed to utilizing the circuitry in the brain that allows them to observe others carefully. They can become emotionally isolated even though surrounded by an adoring public or an always approving professional staff.

Emerging neuroscience about mirror neurons and empathy (Iacoboni, 2008), linked with choice theory and its delivery system reality therapy, will prove increasingly helpful as people learn new ways to connect with each other effectively. Juxtaposing this new and continuously expanding knowledge with the understanding of behavior through the feedback loops of choice theory concepts (Roth & Goldring, 2008) is an exciting prospect for the future. Jointly learning about neuroscience concepts such as mirror neurons provides clients with an additional and unexpected benefit as a way to learn new ideas and an opportunity to apply this new learning together.

A fundamental goal in counseling prominent couples lies in assisting them to learn how to be receptive instead of reactive. Reactive behaviors can also be a defense against perceived external control. Misunderstanding is decreased when partners consciously track their own needs with self-evaluation. Active listening and processing of information reduce negative imaginings. Greater trust and connection come from asking for clarification and speaking truthfully about thoughts and feelings.

For the couple to become happier, each partner must also understand the individual self. Choice theory is based on the principle that human beings can control only their own behavior, but not that of others. Compassion for one another grows from using active observing and listening skills. Recognizing and responding to facial expressions, body postures, and the words and actions of both one's self and others and applying that to one's own behavior can be learned. Justine and Mike learned to check with each other before they automatically reacted to their own perceptions of each other. This became very helpful in improving their relationship and both reported being less reactive and more conscientious as a couple and as parents. Using this process, they both observed and experienced the joy of collaboration.

The main cornerstone of relationship happiness is the ability to control one's own behavior by getting or giving information to others. Four ideas that provide a way to listen and respond actively to each other (Roth & Glasser, 2008; Roth & Goldring, 2008; Roth & Van Der Kar-Levinson, 2002) are:

- Listen actively, watching for facial expressions and emotions.
- Accept the 10% that you may not like or agree with and consider the differences in viewpoints by discussing perceptions, needs, and quality world pictures.
- Speak only for yourself and accept responsibility for your actions rather than blame, criticize, or attempt to control the other person.
- Tell the truth.

CONCLUSION

When couples describe their success in counseling with choice theory ideas, they make comments like: "We are free to talk and laugh together again"; "We focus on the information and make a decision"; "We have a way to make a plan involving each of our viewpoints"; "We were surprised how quickly we began to get along again." The art and science of counseling couples interweaves an understanding of each person and his or her perceptions and needs. Celebrity couples, who come from an even smaller universe, describe feeling expanded. They appreciate the opportunity to improve their closeness and to intensify their connection in a safe environment.

The therapist's goal in couples counseling is to facilitate a happier, more satisfying, and better functioning relationship through increased understanding of one another and positive changes in behavior. Given the additional challenges inherent in working with a celebrity couple, there is a high degree of professional satisfaction when the couple reports that they have developed creative action plans, implemented effective strategies, and improved their relationship.

REFERENCES

Carter, R., Aldridge, S., Page, M., & Parker, S. (2009). *The human brain book.* New York, NY: Dorling Kindersley Limited

Cozolino, L. J. (2002). *The neuroscience of psychotherapy: Building and rebuilding the human brain.* New York, NY: W. W. Norton & Company, Inc.

Glasser, W. (1965). *Reality therapy.* New York, NY: Harper & Row.

Glasser, W. (1998). *Choice theory a new psychology of personal freedom.* New York, NY: HarperCollins.

Glasser, W. (2000). *Counseling with choice theory, the new reality therapy.* New York, NY: HarperCollins.

Glasser, W., & Glasser, C. (2000). *Getting together and staying together: Solving the mystery of marriage.* New York, NY: HarperCollins.

Iacoboni, M. (2008). *Mirroring people: The new science of how we connect with others.* New York, NY: Douglas & McIntyre Ltd.

McGoldrick, M., Gerson, R., & Petry, S. (2008). *Genograms, assessment and intervention* (3rd ed.). New York, NY: W. W. Norton & Company Professional Books.

Roth, B., & Glasser, C. (2008). *Role-play handbook; understanding and teaching the new reality therapy counseling with choice theory through role-play.* Beverly Hills, CA: Association of Ideas Publishing.

Roth, B., & Goldring, C. (2008). *Relationship counseling with choice theory strategies.* Beverly Hills, CA: Association of Ideas Publishing.

Roth, B., & Van Der Kar-Levinson, F. (2002). *Secrets to school success; guiding your child through a joyous learning experience.* Beverly Hills, CA: Association of Ideas Publishing.

Siegel, D. J. (2010). *The mindful therapist: A clinician's guide to mindsight and neural integration.* New York, NY: W. W. Norton & Company Ltd.

Sweeney, M. S. (2009). *The brain, the complete mind, how it develops, how it works, and how to keep it sharp.* Washington, DC: National Geographic Society.

Wubbolding, R. E. (2000). *Reality therapy for the 21st century.* New York, NY: Routledge.

Wubbolding, R. E. (2011). *Reality therapy: Theories of psychotherapy series.* Washington, DC: American Psychological Association.

13

COUPLES COUNSELING AND ILLNESS
The Real Deal

Tammy F. Shaffer

INTRODUCTION

Couples dealing with a chronic illness face many challenges, including the negative impact of the illness on the ability to work, sexual function, mood swings, financial difficulties, and social support systems. Using choice theory and reality therapy, a therapist can help couples look at the impact of chronic illness on getting their basic needs met and how to be more successful in getting needs met despite chronic illness. Reality therapy can have immediate impact, works with a wide variety of issues, and is effective for individuals, families, and couples regardless of age, education, or ethnicity. Reality therapy focuses on the present and helps clients recognize they are more in control of their lives than perhaps they realize; this can be particularly important for those who feel at the mercy of a chronic illness.

The way one sees the world has a major impact on how one feels. Understanding individual perception is the starting place for learning how couples view chronic illness and how perception of the illness impacts all areas of total behavior: actions, thinking, feeling, and physiology. Therapists can help a couple evaluate and adjust their perceptions of their issues, rather than change the issue itself, because that is not always possible. Rather, "the goal of the therapist is sometimes to help clients change their perceptions from painful to less painful and even neutral" (Wubbolding, 2011, p. 51).

PREVALENCE OF CHRONIC ILLNESS AND NEED FOR COUNSELING

The goal of couples counseling is to help couples resolve problems within their relationship. Sometimes, problems are not solvable, such as a chronic illness. According to Bowe, "approximately 54 million Americans (about 1 in 5) have physical, sensory, psychiatric, or cognitive disabilities that interfere with daily living" (as cited in Livneh & Antonak, 2005, p. 12). Therapists can serve an important role in helping individuals and couples learn how to lead satisfying lives despite illness.

Reevaluating oneself and contemplating an uncertain future can shake up even the most emotionally stable individual. Facing a chronic illness alters how people see themselves in their environment. Ideas one has about what his or her world should be like may be challenged. Things important to a client may no longer be attainable or maintainable. If one suffers from crippling arthritis, for example, scrapbooking may become difficult, or downright impossible, because it often requires fine motor control with the hands.

ISSUES IN CHRONIC ILLNESS

Issues that occur most frequently among those with chronic illness include limitations in ability to perform routine tasks, especially those needed to take care of oneself every day; difficulty maintaining employment and in fulfilling role responsibilities as a parent, spouse, employee, or friend; and financial difficulties, particularly as medical bills mount. These issues can result in increased stress for the individual and for the couple's relationship (Livneh & Antonak, 2005). Basic issues that clients with chronic illness face are stress, crisis (with sudden onset of illness), loss and grief, poor self-concept, social stigma (if one's illness can be "seen"), uncertainty about one's future with the illness, and overall quality of life (Livneh & Antonak, 2005).

BASIC NEEDS

Reality therapy focuses on assisting clients in understanding their five basic needs, which Glasser (1998) asserts are genetically encoded and applicable to every human being. These needs are survival, love and belonging, fun, freedom, and power, or inner control. The behaviors we choose are our best attempt to get our needs met—sometimes successfully and sometimes not. Regardless of the effectiveness of our behavior, we continue to evaluate what we are doing to get our needs met and keep behaving in order to gain happiness and well-being.

Each partner can learn how to identify and meet his or her basic needs; this is an important step toward assuming more control over one's life. This is done by assuming responsibility for one's choices and behaviors, which leads to the fulfillment of the needs in a successful manner, as opposed to trying to meet the needs in an ineffective manner, such as alcohol abuse to increase levels of fun, or sex with random partners to meet the need for love and belonging. One also needs to know how to meet his or her needs while respecting others.

Functional Domains and the Basic Needs

Livneh and Antonak (2005) assert that quality of life includes concepts of functional domains such as (a) intrapersonal (e.g., health, perceptions of life satisfaction, feelings of well-being), (b) interpersonal (e.g., family life, social activities), and (c) extrapersonal (e.g., work activities, housing). These domains can be compared to the basic needs discussed in choice theory. Intrapersonal needs may be correlated with the survival need and may overlap several of the other needs. The interpersonal domain may be considered with the love and belonging need in mind, while extrapersonal needs may be linked with power (achievement, success at work). Problems meeting the different domains mean that basic needs are likely to go unfulfilled.

Teaching clients how to find effective ways to meet their needs in spite of a long-term perpetual problem is an important part of the counseling process. If a person is no longer able to perform work duties or is limited in performance, the need for power may be unfulfilled. Power is not the need for control over others; rather, it is the need for success and accomplishment. Financial considerations can also be considered because a potential loss of employment could lead to financial difficulties. Difficulty working and mounting medical costs may also contribute to financial stress and can leave one feeling unable to manage financially or professionally.

Love and Belonging   Depending on the severity of the illness, a couple may experience social isolation as others pull away (Christakis &

Allison, 2006). Glasser (1998) asserted that love and belonging is the most critical of the basic needs. Feeling lonely and alone can have a negative impact on well-being and impacts one's sense of having a place in the world, a sense of fitting in and being an important piece of a puzzle. People need social connections. In fact, a lack of kinship with others can negatively impact one's health, while a sense of connectedness can help enhance health. Therapists do well to help their clients learn how to form social networks. It is important that the couple learn how to meet their need for socialization in ways that consider both their limitations and their potential because "the positive impact of 'supportive social relationship' has gained general acceptance within the scientific community" (Broadhead et al., as cited in Mohr et al., 2003).

Power   Chronic illness impacts both the one who is ill and his or her spouse. Previously assigned family roles may have to be changed, resulting in a shift of power or control (Ellenwood & Jenkins, 2007). For example, the primary financial provider in the family may become unable to work, while the other partner assumes a new role as the breadwinner for the family—assuming more financial responsibility and possibly a greater sense of independence and accomplishment. At the same time, the ill partner is experiencing difficulty meeting his or her need for autonomy and influence and wondering how he or she can still have meaning in life.

QUALITY WORLD

According to choice theory, we all have specific pictures of the things we want in our lives that are seen as important and of value. Glasser (1998) refers to this as our *quality world*. More specifically, the quality world contains: "(1) the *people* we want most to be with, (2) the *things* we most want to own or experience, and (3) the *ideas or systems of belief* that govern much of our behavior" (Glasser, 1998, p. 45).

Teaching Clients How to Identify Their Quality World

To teach clients about the concept of a quality world, the therapist can have them draw pictures of their favorite person, activity, place, and possession. This activity helps clients see and share with their partner what they consider to be valuable and important. Once clients have identified some things that are important to them, they are asked to imagine picking a size for their pictures and placing them in order of importance in an album. This exercise helps clients (1) identify what is important to them, and (2) prioritize these wants.

Sometimes, pictures in one's quality world do not work well together. For example, if one values health and fitness, yet also really likes cheese-cake, adjustments may need to be made. Choices are present and have consequences. Overweight or fitness? Can both be had? Probably not. Perhaps fitness is the 8 × 10 picture, while cheesecake is demoted to a wallet size. Dessert is still in the quality world, but it may now have less importance compared to health.

Changing the Quality World

In working with a couple facing chronic illness, it is important to help the clients reformulate their quality world. Sometimes, a picture in one's quality world may need to be taken out, and this is a challenging process. For example, a person getting a divorce may eventually remove his or her ex-spouse even though the spouse was once an important picture in his or her quality world. In the case of chronic illness, the picture of being healthy may need to be altered to reflect a new perception so that the quality world reflects the client's actual world.

Therapists can help clients examine new methods of fulfilling their needs for power, freedom, love and belonging, survival, and fun. The importance of the needs and how they are met vary from person to person. Therapists can focus on helping clients identify how to fulfill their needs in unique ways, in spite of the challenges clients face regarding chronic illness or other life challenges. A therapist using reality therapy can help people assess how they see the world and how this view impacts their sense of quality of life. Clients may initially resist the idea that how they see the world is relevant because they often see themselves as being at the mercy of their circumstances or others. Therapists using reality therapy help clients review and assess their total behavior in relation to what they want. Total behavior includes thinking, acting, feeling, and physiology. Therapists focus on the acting and thinking because they are the most direct route to change.

According to Lundman and Jansson (2006), individuals can learn to live a good life, even if they have a long-term disease. To do so, they must be able to understand what is going on and have a connection to others. Having a safe haven in which they can be themselves helps to reinforce a positive self-image. People facing a chronic illness, especially one requiring major adaptations, may need to redefine how they see themselves and how they define being "OK."

To form a new definition of "OK," individuals may need to challenge negative self-talk, develop communication skills, establish some new goals, clarify personal values, and determine where time and energy will be spent. As Lundman and Jansson implied, in order to feel better, clients must DO something. "You can't talk your way out of what you've behaved yourself into" (Covey, 1989, p. 186).

ISSUES FACING CAREGIVERS

"Patients ... frequently report significant concerns regarding their loved ones, including strengthening relationships with loved ones, relieving burden on loved ones, and helping others" (Mohr et al., 2003, p. 620). Caregivers, including spouses, experience both emotional and physical impact due to caregiving. Mohr et al. (2003) reported that partners of those with a chronic illness experience higher psychological distress than the partners who are actually sick. The emotional component of their total behavior is also significantly affected; caregiving spouses are almost six times as likely to experience anxiety or depression compared with noncaregivers. The physiological component of total behavior is also impacted: Caregivers developed chronic illnesses at more than twice the rate of peers who did not provide such care (Mohr et al., 2003).

THE UNSOLVABLE PROBLEM

Couples might look for ways to make compromises to help each other through the chronic illness. However, compromises might not always be forthcoming, or possible. Sometimes, "in couple relationships, one partner's self-interests may conflict with the interests of the other" (Rosenblatt & Rieks, 2009, p. 197). If one partner loves to dance and the other no longer has the ability to dance, compromise may not be a real option. Each partner in a relationship can examine his or her needs, how they have been met in the past, and how they might continue to be met.

When needs are not met, individuals may assume ineffective behaviors in a vain attempt to meet their needs. Examples of ineffective behaviors include bossing others, abusing alcohol or other drugs, overeating, fighting, yelling or verbally abusing others, and manipulating others to gain a sense of control. This may be especially true in dealing with couples facing chronic illness, which is, according to Gottman (1999), a *perpetual problem,* or a problem with no solution. "Real relationships have perpetual problems to deal with. For example, differences that they cope with that they don't solve. Most relationship problems don't get solved; they get coped with" (Gottman, as cited in Young, 2005, p. 223). Perpetual problems involve "basic differences in needs that are central to their concepts of who they are as people" (Gottman, 1999, p. 96). Using reality therapy to help a couple explore needs may lead to creative solutions and reducing the problem, even if a solution is not available. For example, instead of dancing, holding hands while listening to music might evoke good moments and memories.

CASE STUDY

Lynn, age 35, and Martin, age 45, were a married couple facing multiple medical crises. They had been married for 15 years and came to counseling to discuss concerns in regard to chronic illness in their family. They had four children, ages 15, 12, 10, and 4. The three older children were males, and the 4-year-old, Anne, their only daughter.

Lynn called herself a "housewife." She saw her responsibility as taking care of her family and home. She had wanted to attend college and become a registered nurse, but put her dream on hold when her first son was born when she was 20 years old. Martin was a plant manager and typically worked more than 70 hours a week. He reported a significant amount of work-related stress, including always being on call.

Lynn and Martin's challenge with chronic illness began with the birth of their daughter. Anne was born with a congenital heart defect and has endured multiple surgeries, medical procedures, and hospital stays. The daily medication regimen for her takes both time and effort, and timing of medications is critical; regardless of where the family may be, medications have to be administered on a strict schedule. The prognosis for Anne was guarded and the couple lived in fear of losing their daughter.

This couple was also dealing with Lynn's health issues. When Anne was 2 years old, Lynn underwent diagnostic tests after having several migraine headaches. She had a brain aneurism that was difficult to fix and surgery was risky. Lynn had significant anxiety about her health and wondered, "What if something happens to me? Who will help Martin raise the kids? They need me!"

The Counseling Process

At the beginning of the interview, Lynn and Martin shared that their challenge with chronic illness began with the birth of their daughter. She was born with a birth defect and required immediate emergency care. In talking about Anne's birth and her emergency transport to another hospital, Lynn became tearful. She also became flushed, avoided eye contact, and looked up at the ceiling while fidgeting with her hands.

Therapist: This seems overwhelming to you. Your face is getting red; you're tearing up, and even wringing your hands. Tell me what you feel inside.

Lynn: My stomach hurts, and I'm getting a headache. ... It's so hard to remember that time. ... My baby was so sick, and I'm her mommy!

Therapist: You must have felt very powerless ...

Lynn: Yes.

Therapist: What were you saying to yourself?

Lynn: My baby needed me, and I couldn't do anything to help her. I also thought it was all my fault. Did I do something wrong?

Commentary The therapist helped Lynn identify that she felt powerless. She was unable to heal her daughter or even fully comprehend the complex medical issues involved. She was also blaming herself for her baby's illness, despite good prenatal care. The only thing Lynn can control is her own thoughts and actions. Blaming herself does not change the situation and is hindering her ability to be present with her child. She can benefit from learning that she can control only her own behaviors and thinking. She cannot control life's circumstances, only her response to them. The discussion then moved to Lynn's health issues, which surfaced when Anne was 2 years old.

Lynn: I had an MRI, and never thought I'd find out that I have an aneurism. It was kind of a shocker. It's easier to deal with my health problems than my child's.

Therapist: How do you take care of yourself through this?

At this point, Lynn begins talking about the impact of chronic illness on her marriage:

Lynn: It affected us a lot. It was to the point we didn't even know if we would last

Martin: Anger ... I was ticked off that we had even more health problems to deal with. I mean, aren't Anne's problems enough? It feels like we're being attacked! We can't go anywhere, or plan anything, because we never know when Anne might get really sick, and she's always going to the doctor. The poor thing can't help it, but it's hard ...

Lynn: Both of us were scared. We didn't know what to expect, and we weren't told a lot of information. We took it out on each other and fought a lot. It affected the whole family, not just me and him.

Commentary Lynn added that her other children were angry, accusing her of abandoning them as she stayed at the hospital with Anne for 6 weeks. The family system was impacted by chronic illness. The need to feel connected was being impacted by the illness and the associated stress. The family was facing a challenge in regard to their need to love and belong. "Siblings of chronically ill children have been found to be at risk of emotional problems, including resentment, aggressive behavior, confused thinking, guilt, embarrassment and feelings of neglect" (Pearce, 2008, p. 224). Adults also experience these feelings and thoughts. At this point, the discussion moved toward social support and spiritual needs.

Lynn: Being able to have God and a church family really helped us. This church that didn't even really know us reached out to us …

Martin: Support from a church mattered so much. … It was the key. … A lot of so-called friends showed their true colors and went out of our lives. Other friends came into our lives. You know who your true friends are when the chips are down … and your friends stay.

Lynn: We didn't go to church our whole marriage until Anne was so sick. Now, we go almost every Sunday …

Commentary Lynn and Martin were fulfilling their need for love and belonging, and a sense of spirituality, by becoming involved in a church. This was a success for this couple. Going to church and being open to making new friends was a choice they made to help fulfill some of their needs. Asking them, "How did you do that?" helped identify the steps they took to become involved in a church and to become a part of a social group. This also helped them evaluate their enjoyment and feelings of happiness at this involvement, leading to increased likelihood of (1) continued church involvement and (2) increased awareness that they were active participants in choosing to increase socialization. Thus, they are not the victims of circumstance.

Lynn and Martin discussed their decreased sense of isolation, after beginning church attendance, and decreased feelings of anger and resentment at friends who had distanced themselves. They chose to focus on new friendships. Martin said their new friends "treat us like there's nothing wrong. We can talk about our problems, but we talk about other stuff, too." They had found a way to meet their need for connection with others and with their God; they now had more ability to enjoy their lives despite circumstances.

Although many may argue that thoughts cannot be controlled, Lynn stated that she learned to discuss topics other than illness and learned that she "can't dwell on it." The truth is that she *could* choose to dwell on an issue; however, she learned to self-evaluate her thoughts and behaviors and determined that endless worrying did not get her any closer to what she wanted, which was feeling peaceful and content and engaged with others.

Therapist: Do you think you've found a new sense of normal, a sense of balance in your life?

Martin: Just like after any major disaster, there's calm after the storm, a new normal. It's OUR normal.

Therapist: What's something you'd recommend to another couple in your situation? Something you do that helps take care of you as a couple?

Lynn: Not to lose hope and not to give up on each other. You can lean on each other and help each other through the grieving process

In the preceding statement, Martin is acknowledging how to make it happen: As opposed to being subject to emotions being thrust upon him, he realized that he could, instead, influence them.

Commentary: Self-Evaluation

It is important to determine what clients can do to respond to upsetting feelings and thoughts because changing the way one thinks and acts is what can lead to feeling better emotionally. Using reality therapy, it is important to move into action after establishing what the clients want, rather than remaining stuck in a potentially endless cycle of feelings.

It is necessary to evaluate what one is doing to get what one wants. The therapist assisted the couple in self-evaluating by asking, "What was helpful? What helped you worked through your loneliness, fear, and anger?" This led to a discussion of what the couple was doing, helping them become more aware that progress is a process—a result of doing something differently, rather than a random result.

In asking the couple what they would recommend to others experiencing chronic stressors in their families, the therapist helped them recognize that instead of remaining victims of their troubles, they had decided to control what they could control, which was their perception of their issues. Martin said, "I have to watch my thoughts," which is a key component of behavioral change in that cognition is a component of total behavior that is subject to alteration more directly than feelings.

Assessment

Using the *WDEP* process (Wubbolding, 2000, 2011), the therapist helped the couple identify what they were unhappy about. This led to identification of basic needs that were not being successfully met. After helping the couple identify their unmet needs, the therapist explored the quality world pictures of what they wanted (*W*) to have in their lives that would satisfy their unmet needs. Their quality world included a picture of a healthy child and a healthy Lynn. These wants were unattainable. Therefore, the therapist worked to help the couple recognize what it would mean to them if they had what they wanted. They both agreed that they wanted relationships with others outside the family.

Following the exploration of what the couple wanted, the therapist helped them examine what they were currently doing (*D*) to get what they wanted. This helped the couple consider their actions and their attempts at getting their wants and needs met. For example, the therapist helped the couple identify that they were lonely and felt isolated.

Evaluation (*E*) of their behaviors followed, to determine the effectiveness of what they were doing.

Finally, a new plan (*P*) would be created so that the couple could find a way to meet some of their needs while still managing all the challenges of the chronic illnesses they faced. The couple self-evaluated by recognizing that they could redirect their thoughts and choose what they wanted to focus on. They also recognized that, despite illness, they still had the choice to act differently. In their case, they chose to attend church and made attempts to establish new friendships, as opposed to being upset and isolating themselves in response to those who left their social circle.

Treatment Goals

The goals for this couple were multifaceted. It was clear that when the couple felt isolated and abandoned, they lacked a feeling of social support. The couple reported a desire to relax and lower the level of stress. They also identified feelings of fear and lack of control over their medical issues and Anne's safety. The therapist helped them determine that their previous efforts to control their situation were ineffective and made plans with the couple to get their needs met in more satisfying ways. The therapist helped the couple to identify ways to help them gain social support, but also taught them how they could get their needs met within their relationship. She helped them make plans for fun and freedom, while also helping them find ways to be sure that Anne was safe in the care of others. They began to recognize what they could control and gained a bit more power when they recognized that they could make some choices about the direction their lives were taking. Lynn and Martin learned how to repeat behaviors and thoughts that effectively led to fulfillment of their basic needs. Identifying successful moments helped them identify a plan to help them continue to meet their needs.

DISCUSSION

It is important that the therapist consider total behavior when working with a couple dealing with a chronic illness. Human behavior, composed of acting thinking, feeling, and physiology, is purposeful and designed to close the gap between what the person wants and what the person perceives that he or she is getting (Glasser, 1998; Wubbolding, 1988, 2000, 2011).

Reality therapy works primarily with thinking and acting because these are the areas in which clients have direct control. However, when working with couples dealing with chronic illness, physiology also demands attention. "The physiology of the body, whether in good health or not, in turn, helps to determine the behaviors we choose" (Uppal, 2003, p. 28). Therapists need to address the influence of physiology, while encouraging the clients to consider that, despite the unwanted impact of physiology, the basic need of power and autonomy can still be met. Despite being ill and perhaps not being in control of the illness, clients can choose how to respond to the sickness. "Every person has internal powers—coping resources which he or she can mobilize in stressful situations to cope more effectively" (Weisler, 2006, p. 38).

One manner in which a couple can respond to sickness or chronic illness is to decide how they can alter their lifestyle to maximize abilities, helping to fulfill the need for autonomy and power as well as love and belonging, freedom, fun, and survival. For example, if a couple is dealing with a partner's difficulty with physical mobility, it may be beneficial to consider alternatives that might help navigate this challenge, even though the challenge may not be resolved or the problem "cured." Can the partner facing the physical challenges use crutches, a wheelchair, or a scooter to enhance ease of movement? Medical personnel, such as physicians and physical therapists, will need to be called upon for their expertise. Therapists may help clients learn how to accept challenges and find alternate ways of fulfilling needs.

SUMMARY

Chronic illness can have lasting effects on relationships, both in the immediate family and within social circles. Without awareness of the challenges related to the illness and the ability to respond effectively to the challenges, relationships may become conflicted and even result in emotional and physical disconnection and divorce.

Fortunately, there is hope. Using reality therapy to help couples take a look at the issues they are facing and helping them recognize they can choose how to handle their problems can help them maintain relationships in a more effective way, thus meeting some of their basic needs. A chronic problem may not have an easy solution or even a solution at all. This does not mean that misery is all the couple has left. By looking at ways their needs of love and belonging, power, fun, survival, and freedom are being affected by the illness, a couple can move to thinking about their goals and how they are, or are not, making progress toward their hopes and dreams. They can self-evaluate by looking at what thoughts and behaviors they have that work, change what is not working, and set themselves up to work as a team for their future. This will be a process, not an event, because life requires continual movement.

REFERENCES

Christakis, N. A., & Allison, P. D. (2006). Mortality after the hospitalization of a spouse. *New England Journal of Medicine, 354,* 719–730.

Covey, S. R. (1989). *The 7 habits of highly effective people.* New York, NY: Simon and Schuster.

Ellenwood, A. E., & Jenkins, J. E. (2007). Unbalancing the effects of chronic illness: Non-traditional family therapy assessment and intervention approach. *American Journal of Family Therapy, 32,* 265–277.

Glasser, W. (1998). *Choice theory: A new psychology of personal freedom.* New York, NY: HarperCollins, Inc.

Gottman, J. (1999). *The marriage clinic: A scientifically based marital therapy.* New York, NY: Norton.

Livneh, H., & Antonak, R. F. (2005). Psychosocial adaptation to chronic illness and disability: A primer for counselors. *Journal of Counseling & Development, 83,* 12–20.

Lundman, B., & Jansson, L. (2007). The meaning of living with a long-term disease. To revalue and be revalued. *Journal of Nursing and Healthcare of Chronic Illness* in association with *Journal of Clinical Nursing 16*(7b), 109–115.

Mohr, D. C., Moran, P. J., Kohn, C., Hart, S., Armstrong, K., Dias, R., Bergsland, E., & Folkman, S. (2003). Couples therapy at end of life. *Psycho-Oncology, 12,* 620–627.

Pearce, F. (2008). Prioritizing needs in the context of chronic illness. *Australian & New Zealand Journal of Family Therapy, 29*(4), 224–225.

Rosenblatt, P. C., & Ricks, S. J. (2009). No compromise: Couples dealing with issues for which they do not see a compromise. *American Journal of Family Therapy, 37,* 196–208.

Uppal, R. (2003). Using reality therapy in the field of physical therapy. *Family Journal, 22*(2), 28–31.

Weisler, S. (2006). Cancer as a turning point in life. *International Journal of Reality Therapy, 26*(1), 38–39.

Wubbolding, R. E. (1988). *Using reality therapy.* New York, NY: Harper & Row.

Wubbolding, R. E. (2000). *Reality therapy for the 21st century.* New York, NY: Routledge.

Wubbolding, R. E. (2011). *Reality therapy: Theories of psychotherapy series.* Washington, DC: American Psychological Association.

Young, M. A. (2005). Creating a confluence: An interview with Susan Johnson and John Gottman. *Family Journal, 13,* 219–225.

III

Proactive Approaches to Good Couple Relationships

14

WHAT WILL IT BE LIKE BEING MARRIED TO ME?

Mark J. Britzman and Sela E. Nagelhout

INTRODUCTION

I am tired and it is my last counseling appointment for the day. Fortunately, I am reenergized because Ray and Cathy are a new referral and I am always up for the challenge of helping strengthen a marriage. I begin the marital counseling process discussing informed consent, including the desired outcome. Suddenly, Cathy looks at Ray and starts yelling,

"I told you we should have done this 6 years ago and you wouldn't listen; now it is too late! I want out!"

"Wow, I am so sorry; I wish I knew how dissatisfied you were. What will happen to our kids?" Ray replies.

"Oh—now you act like you care in front of a counselor? How can I stay married to you? You are just an untrustworthy slime ball."

"I would never have had an affair if you were not such a cold witch," states Ray.

"Ouch!" I am thinking.

"Cathy, uh, Ray—could you try to take a deep breath and let's start the session over?" I said. "I don't want to discount your anger, but I don't think it is helping you get what you want from this session. Is it possible to act as if your marriage could be improved or have you totally made up your mind that you want a divorce?"

Cathy says, "Ask him; it's entirely his fault!"

Ray replies, "Do you see what I have to put up with?"

There must be a better way!

Couples therapy is certainly not boring; however, there is often so much anger and resentment that it is tough to sustain any positive conversations when so many damaging words have been spoken. Seemingly, couples often seek counseling and look to me as if I (Mark Britzman) am a marital judge rather than a counselor. What they really want is to state their contempt in an articulate way and have me cast a verdict of "guilty" on the other partner, thus taking their side and agreeing that their spouse is hopeless. This is obviously not the recipe for reconciliation. I left this session with a headache and wondered, "Is there a better way to ensure that couples do not fall into the abyss of problem saturation and endless blaming and contempt?"

In an attempt to combat all the negativity surrounding marriage and in order to prepare couples with the skills and knowledge required to enter this most sacred and important of relationships proactively, we, the authors of this chapter, have developed *the marital preparation program* (Britzman & Nagelhout, 2011).

Research clearly indicates that a healthy marriage contributes to our health and well-being and is protective of children (Markman, Stanley, & Blumberg, 2010). Therefore, a program that arms couples with the tools to promote marital satisfaction and success is of value.

In a society that puts much significance on the wedding day, we tend to deemphasize preparing for a very important lifetime commitment: marriage. Individuals contemplating marriage often spend quite a bit of time wondering what it will be like to be married to their future spouse. However, in premarital counseling, we encourage our participants to consider their own strengths as well as identify at least one area of personal growth in an attempt to answer the question: "What will it be like to be married to me?" Perhaps even more important is seeking to understand the required ingredients for happiness and need satisfaction of one's future spouse and then developing a plan of action to bring out the best in him or her (Britzman, 2009).

THE CURRENT STATUS OF MARRIAGE IN THE UNITED STATES

We are constantly reminded of the 50% divorce rate in this country; however, that statistic is misleading due to the protective factors that significantly reduce the erosion of marital relationships. It is probably not helpful to look at a new premarital class and wonder which half will divorce. Seemingly, there is much emphasis on divorce rates but not much focus on healthy marriages and factors that significantly reduce the divorce rate. Studies from the National Marriage Project (2009) from the University of Virginia indicate that the look of marriage in America has changed over the last several decades. Today, couples are choosing to marry at an older age or not at all. The number of cohabitating couples is increasing, and more children are being raised in solo-parent households.

This trend has been associated with a number of recent social developments. For instance, in 1969 when Ronald Reagan was governor of California, the state instituted the first no-fault divorce law. The intent of the law was to eliminate strife and deception, but as almost every state adopted a similar law, the divorce rate over the next 20 years almost doubled. President Reagan later remarked that, in retrospect, this law was one of the biggest mistakes of his political career (National Marriage Project, 2009). The divorce revolution peaked in the 1970s, but left a poisonous legacy in which the perception of the institution of marriage was damaged for many. The damage was also felt by many, especially the children of divorce.

The sexual revolution of the 1960s and 1970s contributed to an increase in the incidence and acceptability of extramarital affairs and most likely still affects the state of marriage to some extent today (National Marriage Project, 2009). Additionally, an increased recent emphasis on "individualism" and "self-fulfillment" has appeared to increase the attitude of "what is in it for me?" This sense of entitlement is the breeding ground for selfish behaviors and decreased compassion, respect, and empathy necessary to meet one's spouse's needs. Because we understand the destructive nature of this attitude, our premarital couples, as well as our struggling marital couples, are frequently reminded of the importance of being 100% responsible for their own 50% of the relationship.

Almost everyone has heard the grim 50% forecast for marriage failure in our society. However, research indicates certain variables decrease that figure significantly (see Table 14.1). Based on this information, it is apparent that divorce in our culture is skewed toward poverty and lack of education. It is important for couples getting ready to venture into marriage to understand that their chances for divorce are much lower than they previously have been led to believe. If, in fact, they have some education that provides them with a means to make a decent living,

Table 14.1 Factors That Contribute to Marital Success and Decrease Risk of Divorce

Factors	Percent Decrease in Risk of Divorce
Annual income over $50,000 (vs. under $25,000)	−30
Having a baby 7 months or more after marriage (vs. before marriage)	−24
Marrying over 25 years of age (vs. under 18)	−24
Own family of origin intact (vs. divorced parents)	−14
Religious affiliation (vs. none)	−14
Some college (vs. high school dropout)	−13

Source: The National Marriage Project (2009), p. 80 (http://www.virginia.edu/marriageproject/pdfs/Union_11_25_09.pdf)

have religious beliefs, come from an intact family, and have not already started a family, their chances of success are greatly improved. In our premarital workshops, we emphasize that we are not trying to help our couples learn how "not to get divorced," but instead how to become "masters of marriage," enjoying richly satisfying lifelong unions. We often notice an almost palpable sense of relief in our couples when they are given this encouraging information about their chances for success.

Although marital trends are important to understand, the essence of this chapter is to disseminate specific factors that will likely increase the chance of a sustained and enriched marriage. According to Dr. William Glasser, MD, the founder of choice theory and reality therapy, there is too little focus on making one's partner happy. Jon Carlson, distinguished professor at Governors State University, stated in the forward of *Eight Lessons for a Happier Marriage* that "we each become very clear about what our partner needs to do to improve but seem oblivious of the need to change ourselves" (Glasser & Glasser, 2007, Kindle Section, 1315; Location 97-100).

WHAT DOES AND DOES NOT WORK

External control seems to be a recipe for unhappy marriages. That is, believing that one knows what is best for one's spouse and resorting to criticizing, blaming, complaining, nagging, threatening, and even bribing as a reward and attempt to control will not work in the long run. The seven caring habits that appear to nourish marriages include supporting, encouraging, listening, accepting, trusting, respecting, and negotiating differences, all of which will increase need satisfaction in any marriage (Glasser & Glasser, 2007).

We all have basic needs that, when satisfied, are intimately linked to experiencing an enriched marriage:

- Enriching relationships that promote meeting our need for *love and belonging* by seizing opportunities to bring out the best in others
- Engaging in meaningful activities that fulfill and promote greater feelings of self-worth, significance, and recognition (i.e., *power* or empowerment)
- Having the *freedom* to take control of our lives without coercion, blame, and criticism
- Engaging in enjoyable activities that promote feelings of *fun*
- Making choices that improve our *survival:* physiology, health, and overall sense of wellness (Britzman, 2009; Glasser, 2000; Wubbolding, 2011)

During our workshop, we ask individuals to identify their own top relational needs and also what they believe the top needs of their partners are. When they are finished, we ask them to compare notes to see how they did. Couples enjoy this activity and we believe it is an important one in encouraging them to begin to think in terms of being aware of and satisfying one another's needs, possibly even before their own.

Individuals appear to benefit greatly by asking themselves the following question: "What would it be like to be married to me, and what can I do to bring out the best in my spouse?" It helps to be a "talent scout" consistently, seizing opportunities to share mutual appreciations each day and refraining from negativity. Specifically, keeping love alive requires the following:

- Consistently taking an active interest in one's partner's life
- Treating one's partner with respect—always "seeking first to understand" (Covey, 2004)
- Helping one's partner feel significant and focusing on his or her positive attributes
- Learning to deal with conflict in a gentle, positive way that softens anger and promotes compromise and a sense of teamwork
- Developing hopes and dreams that help one and one's partner together live a life that matters (e.g., "What will our legacy be?)
- Refraining from using criticism that too often leads to defensiveness, contempt, and stonewalling

As humans, we have a burning desire to connect with a significant partner at a deep level. It can be an intensely satisfying experience to share our life with someone who supports and encourages us to lead a fulfilled and desirable lifestyle. To do this successfully, we need to be intimate, which necessitates the safety to become vulnerable to express our true self. If we could only treat our marital partner like our best lifetime friend, life would be so much richer, more stable, and more meaningful. This necessitates a commitment to nourish our marital relationship each and every day, without the fear of being humiliated or rejected. Having intimate, lifelong relationships based on mutual respect, integrity, and love is one of the strongest predictors of sustained happiness and joy.

In fact, research indicates that healthy relationships, and especially enriched marriages, promote well-being. Happily married men and women tend to be healthier, live longer, and experience a perceived high quality of life (Meyers, 2000). Keeping love alive in a marriage can be challenging because of lack of awareness, understanding, and interpersonal development. The goals and objectives of an effective premarital program can increase the likelihood of marital satisfaction. As an example, couples can learn to listen at a deeper level to ensure their partners feel fully heard and understood. Each couple can be helped to discard the perception of "it's you against me." Rather, "it is you and I against the problem." According to Wubbolding, "This mind-set reduces guilt and hostility and also leads to a 'we will work this out' attitude" (2000, p. 82).

THE MARITAL PREPARATION PROGRAM

The marital preparation program designed by Britzman and Nagelhout (2011) includes the goal of helping couples "seize opportunities to bring out the best in one another." This is accomplished via the program's objectives, which entail the following:

- To understand fully the important components that contribute to marital enrichment
- To respect and honor differences influenced by gender, personality, family background, and culture
- To improve communication skills with an emphasis on optimal and consistent listening
- To use encouragement and other loving gestures consistently
- To ensure joint agreement related to budget and charitable giving
- To become aware of relationship strengths and growth areas

The marital preparation program achieves the goal and objectives by developing an educational and experiential model that emphasizes engaging each couple to understand each other's vision for an enriched marriage, to develop an optimal direction to make that vision become a reality, to discover consistent rituals to self-evaluate the marriage, and constantly to develop plans to nourish the marital relationship throughout the couple's lives by ensuring need-satisfying attitudes and behaviors. The program is offered in a variety of differing formats, but given the busy schedules of couples in today's world, the optimal format includes a 6-hour workshop followed by an individual session, meeting with each participating couple to summarize the program's major tenets and ensure a plan to amplify relational strengths.

A Sampling of the Marital Preparation Program's Outline and Activities

The WDEP system, developed by Robert Wubbolding, EdD, director of the Center for Reality Therapy, develops a constructive atmosphere and useful process to organize the marital preparation program. Meaningful and engaging couple activities include:

- Exploring wants: What is your vision for an enriched marriage? What do you want that you are not getting? What are you getting that you do not want? What are the priorities related to what you want? What do you have to give up to get what you want? How much effort or energy are you willing to exert to get what you want?
- Direction or doing: What are you doing to meet your future spouse's needs? Provide a positive example of when you were getting what you wanted. What direction are your choices taking you in your quest to ensure an enriched marriage?
- Evaluation of self: Is the overall direction in your choices and life's direction the best for you and your marriage? Are your present choices bringing your future spouse closer to you or negatively impacting your relationship? Does it help your relationship to look at things as you currently do? If you could change any attitude or behavior, what would produce the most positive impact on your current relationship?
- Plan: What is possible that you could do to improve your relationship with your future spouse? What would you like to do? What will you do? Is your plan simple, attainable, measurable, and generated by you, and does it ensure a commitment to follow-through? (Wubbolding, 2000, 2011).

The marital preparation program includes, but is not limited to, the following activities linked to the WDEP system:

1. An ice-breaker that asks each participant to articulate what initially attracted him or her to the future spouse. A follow-up question is then asked of each participant regarding the following: "What would it be like be married to you?" (e.g., three positive traits or characteristics, and one area in need of growth and development)

2. Questions related to research findings regarding factors that contribute to or erode marital enrichment are answered by "true" or "false."

3. Each couple tries to guess the "Four Horsemen of the Apocalypse" out of a list of 10 possible factors, all with the potential to be related negatively to marital enrichment (Gottman, 1999). An overview and explanation of the "Four Horsemen of the Apocalypse" is given, including linkage to *Seven Deadly Habits of Marriage* (Glasser & Glasser, 2007).

4. Each couple is then asked to rank his or her partner's top three needs in marriage out of a list of 10 possibilities. This helps each to become more familiar with the quality world regarding relational needs.

5. A brief summary of connecting habits is then disseminated to each couple with an opportunity to link to their current relationship (Glasser & Glasser, 2007; Gottman, 1999).

6. Each couple is then taught optimal skills related to listening, understanding, and conflict resolution; this is stimulated by a myriad of statements to be completed (see Table 14.2).

7. Each couple is also given a Myers-Briggs type indicator (CPP, Inc., Self-Scorable Form M, 2011) followed by an activity to promote respect and further discussions related to differing temperaments and preferences.

8. A discussion is initiated regarding future plans and opportunities to create an optimal monthly budget to improve financial management.

9. Each partner is then encouraged to write his or her own vow with an emphasis on an ongoing vision and commitment for bringing out the best in one another.

10. A follow-up meeting is then scheduled with each couple to review the program and disseminate results from the premarital assessment that is currently being factor analyzed (Britzman, 2011).

Table 14.2 Statements Used for Couples to Improve Listening Skills and Stimulate Deeper Understanding

The aspect(s) I most appreciated about my parent's marriage was/were …

The aspects(s) I least appreciated about my parents' marriage was/were …

To be totally accepted by my family, it seemed like I had to …

If you interviewed my siblings, they would describe me as …

I really enjoy spending time with you when we are doing …

I believe we are compatible to live the rest of our lives together because …

My only anxiety about you being married to me is …

Something I have not fully disclosed to you about me is …

I know our biggest differences relate to …

My biggest insecurity about you being married to me is …

What I have learned that I want and do not want related to past relationships is …

I know I choose to get angry when …

I will likely have difficulty compromising on the following …

I believe the best conflict resolution process that would work for us entails …

I am willing to show my love for you in the following nonsexual behaviors …

I will signal to you that I am interested in making love by …

I will communicate what I want and do not want sexually by …

My vision of a great marriage and family life includes …

I propose that we consistently evaluate the overall health of our marriage by …

I would know if our marriage is suffering if …

I believe a good plan to be reminded of our marital vows and to nourish our relationship needs to include …

SUMMARY AND CONCLUDING DISCUSSION

A great deal of research concerning marriage has been done in the last few decades and has revealed a large body of knowledge that was previously speculative at best. We now understand what social forces enhance and erode the marital union as well as the personal behaviors and communication patterns that tend to attract or repel our marital partners. Certain personal needs are universally desired and must be met for an individual to experience a deeply satisfying existence. Being part of a loving and supportive marriage can greatly increase the chances of each partner having his or her needs met and ultimately living a longer, healthier, and more meaningful life. Equally important, we now know that improved relational skills can be learned and marital enrichment programs that focus on those skills have been found to be effective in increasing marital satisfaction.

Choice theory provides a conceptual framework that permeates our entire marital preparation program. Ultimately, couples need to be aware and choose attitudes that lead to a positive pattern of behaviors. This includes, but is not limited to, better understanding one's partner's perception of his or her ideal vision of marriage, becoming intentional to implement connecting habits on a consistent basis, and developing rituals to self-evaluate marital satisfaction. It is also important to ensure there is an ongoing, yet dynamic plan to replace the deadly habits that can thwart a relationship with connecting positive habits in order to "seize opportunities to bring out the best in one another." Consequently, a positive response would be acknowledged when asked, "What would it be like to be married to me?"

REFERENCES

Britzman, M. J. (2009). *Pursuing the good life*. Bloomington, IN: Unlimited Publishing.

Britzman, M. J., & Nagelhout, S. (2011). *The marital preparation program*. Unpublished manuscript.

Britzman, M. J. (2011). Premarital assessment inventory. Retrieved from http://www.pursuingthegoodlife.com/premarital/default.asp

Covey, S. R. (2004). *The seven habits of highly effective people*. New York, NY: Simon & Schuster.

Glasser, W. (2000). *Counseling with choice theory: The new reality therapy*. New York, NY: HarperCollins.

Glasser, W., & Glasser, C. (2007). *Eight lessons for a happier marriage*. HarperCollins e-books.

Gottman, J. (1999). *The marriage clinic: A scientifically based marital therapy*. New York, NY: W. W. Norton & Company, Inc.

Jakubowski, S. F., Milne, E. P., Brunner, H., & Miller, R. B. (2004). A review of empirically supported marital enrichment programs. *Family Relations, 53*, 528–536.

Markman, H. J., Stanley, S. M., & Blumberg, S. L. (2010). *Fighting for your marriage*. San Francisco, CA: Jossey–Bass.

Meyers, D. G. (2000). *The American paradox*. New Haven, CT: Yale University Press.

National Marriage Project (2009). *The state of our unions: Marriage in 2009*. Retrieved from http://www.virginia.edu/marriageproject/pdfs/Union_11_25_09.pdf

Wubbolding, R. E. (2000). *Reality therapy for the 21st century*. Philadelphia, PA: American Psychological Association.

Wubbolding, R. E. (2011). *Reality therapy*. Washington, DC: American Psychological Association.

15

A CONTEXTUAL APPROACH
TO RELATIONSHIP ASSESSMENT

Jeri L. Crowell and Jerry A. Mobley

INTRODUCTION

Couples therapy with reality therapy/choice theory does not occur in a vacuum; the practice of therapy occurs in a series of interlocking systems (Wubbolding, 2000, 2011). While the counseling is decidedly reality therapy, the process for identifying significant issues in the assessment phase often overlaps with intervention. Through exploration of the inseparable elements of total behavior (acting, thinking, feeling, and physiology), reality therapy is applied in an attempt to understand the couple's relationship. Along with the holistic perspective of ecological counseling, the assessment process offers insight into both the context and meaning within the couple's interactions. As the couple implements their combined quality world pictures of what is important in their lives (Glasser, 1998, 2000; Wubbolding, 2000, 2011), reality therapy/choice theory helps us to understand the level of congruence within their individual and combined quality worlds.

ECOLOGICAL COUNSELING AND SYSTEMS
WITH REALITY THERAPY/CHOICE THEORY

The idea of congruence is especially important in both ecological and family systems counseling. In ecological counseling, the term *concordance* correspondingly describes how well individuals thrive within the context of their lives (Conyne & Cook, 2004; Crowell, 2007). Similarly, systems theory focuses on synergy and balance between the separate elements in order to function well as a whole, or balancing individual needs with system needs (Goldenberg & Goldenberg, 2008). Congruence in reality therapy, as described by Wubbolding (2000, 2011), is the intersection of the individuals' wants, or pictures in their quality worlds, and their behavior, or how they live out their quality worlds, in tandem with the fact that two individuals are both attempting to gain satisfaction as a couple at the same time.

The paradigm for human behavior in ecological counseling is that "behavior is a function of persons interacting within their environments" (Conyne & Cook, 2004, p. 6). In these interactions lies the process of making meaning of their world, to which the couple attributes an assessment of value, either positive or negative, on its functioning as a couple. Individuals make sense of their world and develop certain expectations about how things work through feedback from the external world, thus creating subjective perceptions. Wubbolding (2011) described perceptual filters, one of which applies value to incoming information, so that appropriate behaviors are chosen in response to life events. Both the ecological counseling model and reality therapy inform assessment and practice by enhancing a counselor's perspective on the purpose of the couple's behavior and congruence levels between what they want and what they are experiencing within their reality.

In a trusting relationship with a reality therapist, a couple will gain awareness of their interaction patterns in order to develop increased correspondence between what they want and how they plan to achieve satisfaction of their wants within their particular ecology. Couples are engaged in an interdependent relationship where, as Conyne and Cook (2004) state, "a change in any one part of a system will ultimately affect other parts of the system" (p. 29). Family systems theory assesses specific aspects of functioning, such as how the couple have fun together or how they argue and what they argue about. Application of choice theory through reality therapy provides a way of connecting with people who are frustrated with what they perceive they have or what they want and are not getting. Though all behavior is purposeful, all behavior is not necessarily effective.

ASSESSMENT IN COUPLES THERAPY

Couples therapy involves the meeting of two "tectonic plates" from two disparate histories and their respective quality worlds to form a single new "geographic feature" called a couple, marriage, partnership, or significant relationship. Tectonic plates around the earth collide, causing volcanoes and creating mountain ranges, and pulling other earthly spaces apart, which is often what a failing relationship feels like. According to Glasser (1998), "There are only two ways people move away from each other: they resist or withdraw, fight or flee" (p. 173). Family histories with addiction (Fisher & Harrison, 2005; Threadway, 1989) or violence (Gil, 1988; Kemp, 1998) often foment a fantasy-filled quality world of living happily ever after while ignoring the missing relational skill set or self-management necessary to achieve meaningful interaction.

Wubbolding (2000, 2011) explains how people choose behaviors based upon their perceptions of how satisfying or discrepant their lives are in comparison to their ideals, their quality world. Specific behaviors arise out of the discrepancies and are experienced as elements of acting, thinking, feeling, and physiology (Glasser, 1998; Wubbolding, 2011). All four parts of total behavior may be occurring at the same moment, but the intensity of any one part may be more prominent. For example, when the couple meets, the feeling element is often very intense, but as time goes on this feeling element may become overshadowed by thinking in one or both of the partners, which becomes ever more complex. "All you need is love," which is strongest in the beginning of a relationship, often fades rapidly in the light of more realistic assumptions about life, including the use of power to negotiate relationship topics (Napier, 1990).

Assessment that is grounded ecologically examines both the objective qualities of the environment and the individual's subjective perception of those qualities (Wilson, 2004). Contextual data and a systemic perspective enhance the assessment of present and historical influences in order to help a couple discover how meaning-making impacts their separate and combined experiences. Data collection capturing the couple's interactions can be done through genograms, ecomaps, lifelines, self-report questionnaires, and an ecological checklist, such as the one provided in Appendix 15.1. Wilson (2004) has offered numerous ideas for assessment tools to test the goodness of fit in interpersonal interactions, as well as between person and environment. Goodness of fit is used to describe to what degree a person is able to thrive within his or her environment, or the dynamic balance of person–environment interactions. Ecological assessment involves a thorough understanding of assessment methods and instruments available while, as Wilson (2004) emphasizes, maintaining sensitivity to the "social and cultural realities and circumstances of the client's ecological niche" (p. 149).

CASE STUDY

Imagine a couple with these issues: addiction in his family tree with his own marijuana use since age 17 coupled with cocaine use beginning in his 20s, and verbal and physical violence in her family of origin until she left home, pregnant, at age 16. Greg and Jeanene are both divorced with children. He is drug free and working a program of recovery; she has an associate's degree and a responsible job. They found each other in their early 30s.

Individual genograms are used to assess the couple's relationship experiences and family constitutions, while defining what they have been trying to create in their combined quality world. Couples may have a better sense of what they want from the relationship than what they are bringing to the relationship, particularly evident with couples who have addiction or abuse issues. Dysfunctional patterns in the family tree can create a mixture of contradictory wants in a person's quality world. For instance, someone raised in a chemically dependent family has experienced a pattern of inconsistent family behaviors. As a result, a child often grows up with conflicted wants in his or her quality world, such as that the child wants love and belonging, but not when the family hugs one time and hits another.

Challenges/Problems

Jeanene and Greg are afraid to believe that their individual and combined quality worlds have meaning. With only a couple years of sobriety since the age of 16, Greg has minimally begun to emotionally complete his adolescence. In spite of raising children, he is limited in his ability to make commitments and live them out on a daily basis. He is more inclined to initiate plans than to complete them, and the activity that is the most interesting to him, playing baseball, is no longer a realistic goal. Even though he had success playing in the past, he no longer has the quality of health or physical fitness to maintain his athleticism at a competitive level. What can he contribute to the couple's picture of its future and goals when his personal picture is limited?

True to the systemic concept that "water seeks its own level" of differentiation (Bowen, 1978), Jeanene's reality as it relates to her quality world is not any better. She tends to adjust her definition of what she wants constantly, from a particular employment setting, to the marriage, to a certain geographical location, to an item purchase, or to the resolution of a problem, and then she quickly loses interest. Through a life of constant upheaval, Jeanene has learned that a larger unified perspective is not favorable over survival modules that she can focus on and accomplish in smaller doses. Again, the issue can be raised: What can she contribute to the couple's picture of its future and goals when her personal picture is limited?

After years of parenting children as single parents, this couple met and married within months. Jeanene has been the more active mate behind solidifying the relationship; Greg has been responsive to her overtures. However, they established communication patterns of "your kids" and "my kids," indicating interpersonal distrust. For 7 weeks "everything is wonderful" between the couple until Jeanene is terminated from an unsatisfying job for which she also was not well qualified. Her notion is to move the family to another state for better employment. After finding suitable work 50 miles away, Jeanene makes plans to move out of their house, regardless of the newly formed family unit. Greg is wounded by her insistence on moving to the new area to be close to her employment and asks her to commute for 3 months until the end of the school year because of the children and their housing lease. When she moves out and takes her children, he considers the marriage ended. Jeanene and Greg have sporadic contact with each other until he is able to take care of the lease and possibly move. At this point, Jeanene changes her mind about what she has done and what is important.

Case Conceptualization

The clash of visions for Jeanene and Greg is further complicated by the personal and family dysfunctions (see Figure 15.1) that impair the couple's combined quality world. Individual work needs to be accomplished through his inpatient addiction counseling, outpatient support from 12-step work and perhaps a spiritual program, and her ongoing work on family-of-origin issues, before the couple has much to contribute to the third entity of their "coupleship."

In a genogram, the females are shown as circles and males are shown as squares. Both Jeanene and Greg are from divorced parents, as seen by the vertical double lines through the horizontal lines that connect their mothers to their fathers. Both Jeanene and Greg are divorced, as can be seen by the same diagramming. Jeanene's father was violent, as shown by the zigzag line between him and both the mother and Jeanene. In the genogram, a zigzag line goes from Jeanene to her ex-husband, which indicates violence in their relationship. Finally, there is a single vertical

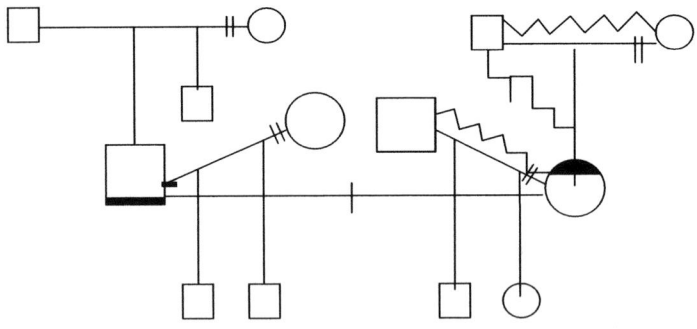

Figure 15.1 Genogram of Jeanene's and Greg's families showing divorces, Greg's addiction, Jeanene's depression, her father's violence against Jeanene and her mother, her first husband's violence against her, and the couple's separation.

line in the horizontal line between Jeanene and Greg because they are separated (not divorced).

The context within which Greg and Jeanene began their relationship was fairly stable financially and emotionally, but once Jeanene's job loss hit the family unit, it became clear that necessary relationship skills for collaboration and cooperation were not developed for the couple. As previously suggested, the important aspect of commitment was perceived as less significant a motivator than the desire for immediate gratification. Jeanene was more focused on the meaning that having a job, even an unsatisfying one, meant to her sense of stability and perhaps her sense of personal power and freedom.

Other data obtained in counseling included more ineffective family dynamics. Jeanene reported that her mother reacted to her father's abuse rigidly with excessive control over Jeanene, overcompensating for her husband in order to see herself as a loving parent. Greg described his father as placating toward a demanding mother, who eventually abandoned the family. The pattern Jeanene and Greg had pieced together for themselves in the beginning was that the woman dominates and the man goes along and adapts. Their individual histories created perceptions that impacted their total behaviors as a couple. The pattern broke down in this case study when Jeanene demanded the move to another town to be closer to her work, and Greg did not adapt well.

Counseling Process

The practice of reality therapy requires the counselor to create an environment of trust so that clients can develop a relationship with the counselor that could model relationships in other areas of their lives.

Important elements of the environment include fairness, friendliness, and firmness, as well as safety and empathy (Wubbolding, 2000, 2011). Greg appeared to be determined to establish a trusting counseling relationship, though emotions ran high at first. Greg was emphatic: "I don't want to open my heart and have it torn out again." Jeanene was contrite and initially took responsibility for moving out and causing the family disruption, but also was quick to rationalize external influences as causal to her choices.

Choice theory stresses that an unsatisfying relationship is at the root of most unhappiness (Glasser, 1998). In initial sessions, reality therapy utilizes a process of investigating the issues for counseling through obtaining answers to questions aimed at perceptions of why counseling is needed. Choice theory provides the theoretical underpinning to conceptualize those counseling issues and the analysis of behaviors, also informed by the ecological assessment of contextual factors. In collaboration with a couple, reality therapists stress the importance of each individual's willingness to change and provide safe environments to explore their quality worlds (Glasser, 2000). Reality therapists explain that our quality worlds provide the motivation for behavior and "the only person we can control is ourselves" (Glasser, 2000, p. 39). Therefore, holistic assessment includes understanding total behavior (acting, thinking, feeling, and physiology), which is how we communicate with the rest of the world (Wubbolding, 2011).

For weeks, Jeanene and Greg discuss with the counselor what they are trying to create in their combined quality world. In addition to what they do not want to happen in the relationship, they are able to state what they do want: mutual sharing; closeness, including sex and mutual physical and emotional support; stability (i.e., "no surprises"); commitment that transcends immediate circumstances; and support for the children. Utilizing their genogram, the counselor assists Jeanene and Greg in understanding the faulty perceptions about relationships that they learned in their families and acted out in their previous relationships.

Not long into the counseling relationship, Greg decided that he needed to stop being passive and become more proactive in the setting of family goals and Jeanene needed to become more comfortable with sharing those responsibilities. Merely accepting what Jeanene wanted to do was not adequate to satisfying Greg's picture of a quality relationship. Utilizing Glasser's (1998, 2000) seven deadly habits and seven caring habits, the counselor worked with Jeanene and Greg together to help them to communicate as honestly as possible using these practices to help undo behaviors that were part of previous interaction patterns.

As a form of assessment, Glasser's habits are used to talk about specific ways of being with each other and then the specific outcomes they had each experienced, wanted to experience, and how these habits could help them get what they wanted.

At the same time, Jeanene realized that demanding what she wanted did not achieve her goal of a long-term, mutually satisfying partnership. In the safety of the counseling relationship Jeanene became aware of the extent of her use of external control psychology and how her behaviors caused anxiety in herself and Greg. The possibility of changing her behaviors shifted the relationship dynamics, particularly with regard to the perception of power. With the assistance of a mild antidepressant to moderate Jeanene's emotional extremes, she has begun to discuss and not demand, and to talk over issues and not tell her husband what was going to be done. Within weeks, Jeanene was commuting to be with Greg for weekends, though Greg was cautiously receptive: "She is saying and doing the right things, but she said and did the right things when we got married," he said.

Meanwhile, counseling included teaching the couple Glasser's solving circle (1998, pp. 94–95) in order to give them a tool for problem solving on their own. The couple read Glasser's (2000) book, *Getting Together and Staying Together: Solving the Mystery of Marriage,* and discussed the book's ideas in counseling sessions. After several months, Jeanene left her job and moved to be with Greg.

Outcome

The couple has been back together for a month. Jeanene and Greg spend considerable time discussing what they are doing before, during, and after they do it. Both seem to be fragile, although Greg states, "We are enjoying each other again, like when we first got together." They maintain contact with their counselor and support people, who have become extended family of sorts, demonstrating care for the couple who became committed to changing extraordinarily unhealthy and destructive behaviors. Ecological counseling and family systems theory would state that just as a change in one individual creates reciprocal change in the other individual within a family, a change in the environment impacts the perceptions of the individuals within that environment.

DISCUSSION

Reality therapy combined with an ecological perspective—and engaging family systems theory in couples counseling, in particular—offers holistic assessment for successful work. Ecological principles and a systemic perspective may be utilized to create an understanding of the present and historical contextual influences that occur on many levels in order to help couples discover how meaning-making impacts their separate and together experiences. Melito (2006) states that the parts of family systems—from subgroups (e.g., parents, siblings) to individuals' intrapsychic processes—need "to be separate and distinct, yet interconnected and able to act in concert" (p. 346). Meaning-making is an invaluable contribution from ecological counseling to the conceptualization of therapeutic interventions. Assessing the quality world to determine the desired outcome, the couple can use their power to adapt their behaviors and increase self-awareness in counseling. Meanwhile, the couple has the opportunity to analyze and understand their family history and have more choices in their individual adaptations or behaviors.

Ecological principles in a systemic perspective enhance the choice theory lens to include understanding the present and historical contextual influences that occur on many levels. Ecological assessment discovers how meaning-making is derived from a couple's separate and tethered experiences and explores the interrelated elements of families, finances, and even fun (Conyne & Cook, 2004; Crowell, 2007). The processes of reality therapy, combined with the information gained from ecological assessment, can help individuals to become more intentional in their choices to achieve the desired result. Assessing the combined quality world aids the couple to gain awareness of historic family approaches because, without this awareness, assumptions can become essentials and produce a clash of patterns.

Glasser's (1998) solving circle gives a couple an understandable tool that can be used anytime and anywhere to solve problems. Along with more traditional tools like genograms, counselors can help the couple define their relationship by assessing their combined quality world to develop a combined sense of what they want from the relationship. This process begins with what they are bringing to the relationship.

Through the Lens of Choice Theory

Counseling with couples focuses on improving their relationship. Meeting human needs both individually and as a couple is a powerful experience and is the result of effective behavioral choices. The complexity of the relationship between just two individuals requires assessment and analysis of their individual quality worlds and then the combined quality world that emerges from the partnership. Reality therapists help family members change behavior, improve communication, and use "quality time" (Wubbolding, 1999)—time that builds positive perceptions of each other and of the world around them. Assessment of goals is the starting process for any couple to determine what they want from the therapeutic experience. In turn, the counselor assists the couple in learning self-evaluation skills to emphasize individual responsibility for behaviors.

Strengths and Challenges of Using CT/RT Working With This Couple

Choice theory (CT) and reality therapy (RT) stress that people are able to create behaviors in response to their perceptions of what is happening in the real world. Jeanene and Greg reacted to each other and to their environment as many people do, based upon their typical behavior patterns. People do things without realizing that every behavior is a choice and an attempt to meet an unsatisfied need. This couple was taught to gain awareness of their basic needs, and their quality worlds. Their combined quality world reflected generations of issues that were available from compiling a three-generational genogram (McGoldrick & Gerson, 1985). In this case, the couple is presented over an extended period of time, which allowed them to develop individually and as a couple. In counseling, they individually clarified and accomplished goals and brought those successes to their "coupleship."

Counselors who know CT/RT and 12-step programs can apply the reality therapy process to a relationship to help the couple accomplish what they want. Problems are usually multifaceted and difficult, and treatment often needs to be equally rigorous. The presenting problem in this case study was not about who would live where, but about unmet needs that represented an ongoing cycle of ineffective behaviors. The counselor helped the couple to evaluate the frustration signals that needs were not met in order to choose more effective behaviors. Similarly, an ecological assessment highlighted the choices the couple had in order to adapt to significant life challenges.

IMPLICATIONS FOR TRAINING AND SUPERVISION

Assessment is critical in counseling, initially to garner enough information so that the counselor can facilitate a couple's desire to work on their relationship. In determining the appropriate questions for discovery of such information, an ecological assessment includes contextual factors in developing an understanding of the problem. Reality therapy emphasizes the relationship people want with others as the integral component of happiness (Glasser, 1998). People choose their behaviors and have more personal control than they realize. It is particularly helpful to identify fantasy within the couple's thought behaviors, as well as where external control behaviors are practiced more than internal control behaviors.

Experience demonstrates that counselors in training benefit greatly from supervision with a strong emphasis on assessment of need strength, degree of need satisfaction, and planning for improvement. As Glasser (1998) states, "There is a lot of security in a plan; there's a sense of control, it's what you can do" (p. 172). Numerous tools are available for training and practice, such as the process of the WDEP, ecological checklist (Appendix 15.1), genograms, and more. It is not enough to be familiar with these tools; their effective and competent use often requires training, consultation, and supervision.

REFERENCES

Bowen, M. (1978). *Family therapy in clinical practice.* New York, NY: Aronson.

Conyne, R. K., & Cook, E. P. (2004). *Ecological counseling: An innovative approach to conceptualizing person–environment interaction.* Alexandria, VA: American Counseling Association.

Crowell, J. L. (2007). *Exploring the ecological counseling model with teachers in an urban school.* Doctoral dissertation. Retrieved from http://www.ohiolink.edu/etd

Fisher, G., & Harrison, T. (2005). *Substance abuse information for school counselors, social workers, therapists, and counselors.* Boston, MA: Pearson Education.

Gil, E. (1988). *Treatment of adult survivors of childhood abuse.* Walnut Creek, CA: Launch Press.

Glasser, W. (1998). *Choice theory: A new psychology of personal freedom.* New York, NY: HarperCollins.

Glasser, W. (2000). *Counseling with choice theory: The new reality therapy.* New York, NY: HarperCollins.

Goldenberg, H., & Goldenberg, I. (2008). *Family therapy: An overview* (7th ed.). Belmont, CA: Thomson Brooks/Cole.

Kemp, A. (1998). *Abuse in the family: An introduction.* Pacific Grove, CA: Brooks/Cole Publishing Company.

McGoldrick, M., & Gerson, R. (1985). *Genogram in family assessments.* New York, NY: Norton.

Melito, R. (2006). Integrating individual and family therapies: A structural-developmental approach. *Journal of Psychotherapy Integration, 16*(3), 346–381. doi:10.1037/1053-0479.16.3.346

Napier, A. (1990). *The fragile bond: In search of an equal, intimate and enduring marriage.* New York, NY: Harper Perennial.

Threadway, D. (1989). *Before it's too late: Working with substance abuse in the family.* New York, NY: Norton.

Wilson, F. R. (2004). Ecological psychotherapy. In R. Conyne & E. Cook (Eds.), *Ecological counseling: An innovative approach to conceptualizing person–environment interaction.* Alexandria, VA: American Counseling Association.

Wubbolding, R. E. (1999). Creating intimacy through reality therapy. In J. Carlson & L. Sperry (Eds.), *The intimate couple.* New York, NY: Routledge.

Wubbolding, R. E. (2000). *Reality therapy for the 21st century.* New York, NY: Routledge.

Wubbolding, R. E. (2011). *Reality therapy: Theories of psychotherapy series.* Washington, DC: American Psychological Association.

APPENDIX 15.1: ECOLOGICAL CHECKLIST

In my assessment of the client(s) before me, I have considered:

____ gender, race, national origin, socioeconomic status, and education level (particularly in selecting and administering any assessment tools)

____ any disability, including language and physical, mental, or emotional handicap

____ the family system: the makeup of the immediate family, relationships of its members, cultural influences, values and morals, and the effects of these on behavior

____ the social network: the makeup and level of support or influence, peer group influences on the client and that the client exerts on others

____ roles within the family, workplace, school, church, and community: expectations and behaviors associated with positions in society

____ psychological, economic, and emotional impact of societal factors, such as opportunity, security, historical traditions, behavioral sanctions, etc.

____ the ways in which the client makes meaning of his or her life experiences and environment, including image of self in context with his or her world

Source: Crowell, J. L. (2007). *Exploring the ecological counseling model with teachers in an urban school.* Doctoral dissertation. Retrieved from http://www.ohiolink.edu/etd

16

USING CHOICE THEORY AND REALITY THERAPY IN PREMARITAL COUNSELING

Sylinda Gilchrist Banks

INTRODUCTION

A common phrase spoken at weddings is "the two shall become one." However, couples often have difficulty merging two sets of ideas, dreams, and desires into one. A couple enters marriage with different personalities, values and needs, expectations, and goals. Creating this new life together and negotiating the differences can be a challenge for any couple. In an effort to deal with these differences, couples enter premarital counseling to gain insight on how to develop a fulfilling relationship. Through premarital counseling, a couple can explore their dreams, their fears, and their differences and come to a greater understanding of their choices, needs, and behaviors.

While many couples are getting married, many marriages are ending in divorce. In 2009, there were 5.3 million marriages and 3.4 million divorces (National Center for Health Statistics, 2010)—a much higher percentage than the 50% divorce rate that is usually suggested in the media. To avoid becoming a statistic, many engaged couples are deciding to become more prepared to face the realities and challenges of marriage through premarital counseling, which has been shown to have a positive effect on marital quality and can prevent divorce (Owen,

Rhoades, Stanley, & Markman, 2011). Moss (1988) reported the top five reasons lawyers cited for divorce (in order of importance):

- Communication
- Divergent personal growth patterns
- Sex, adultery, or lack of affection
- Money problems
- Lack of understanding of marital expectations

Premarital counseling can help couples strengthen their communication skills, establish mutual goals, and resolve conflicts peacefully, as well as establish realistic goals for a happy relationship. Through participation in this type of counseling, couples can learn strategies to avoid the pitfalls of marital discord.

Effective communication is important to any relationship, but especially in marriage. Researchers have determined that negative communication is significantly associated with divorce (Markman, Rhoades, Stanley, Ragan, & Whitton, 2010). With effective communication skills, a couple can address and resolve different issues that occur during the marriage. On the other hand, ineffective communication or the lack of communication will leave a couple without the foundation needed to overcome marital troubles. Premarital counseling can teach couples to develop a deeper understanding of each individual's values, expectations, and perceptions of marriage, and can teach couples how to negotiate differences. Couples with strong communication skills are equipped to articulate their wants, aspirations, and concerns.

Premarital counseling can be provided using various frameworks. However, choice theory and its delivery system, reality therapy, comprise a therapeutic framework to teach couples how to strengthen their relationship. "Choice Theory is useful, even vital, well before marriage" (Glasser, 1998, p. 165). Choice theory and reality therapy teach couples how to understand external control, how to communicate their wants and desires, and how to satisfy basic needs, resolve conflicts, and develop goals.

CASE STUDY

Kimberly and Desmond were planning to get married in 10 months. Their minister recommended they attend premarital counseling. This suggestion is made to all engaged couples planning to wed in the church. Desmond and Kimberly met in graduate school. After 3 years of dating, Desmond proposed on Valentine's Day. Kimberly decided to attend a local university and live with her parents. She works as a physical therapist. Kimberly shares a close relationship with her parents, especially her mother. Every Saturday, Kimberly and her mother enjoy a day shopping, going to lunch, or just completing errands. She has a younger brother, who attends an out-of-state university. Kimberly's parents are excited about the wedding and treat Desmond like a son.

Desmond works for an accounting firm and resides in a home purchased a year ago. Kimberly was involved in the selection of the house but will not move in until after the wedding. Desmond enjoys a close relationship with his parents, even though they were divorced when he was 11 years old. He finds playing basketball and tennis on weekends with his friends need satisfying.

Relationship Assessment

When couples participate in premarital counseling, normally they do not have specific or observable problems. The goal is to enhance the couple's relationship by teaching strategies they can implement after the ceremony. The number of sessions in premarital counseling can vary from one to six depending on the setting. Premarital reality therapy counseling, based on choice theory, is designed to help the couple share their wants or pictures in their quality world, learn to negotiate differences, and learn to communicate using caring behaviors. The WDEP (wants, doing/direction, evaluation, and planning) process (Wubbolding, 2000) is used to give couples a process for making effective relationship choices.

Kimberly and Desmond's goals for counseling were to strengthen their relationship and to identify ways to resolve conflicts and maintain good communication. They did not have any major concerns about marriage and viewed premarital counseling as being proactive.

Identifying Goals

Premarital counseling involves the process of educating and coaching. The reality therapist will educate the couple on the concepts of choice theory and reality therapy. Couples will learn about the quality world, communicating using caring habits, developing a need-satisfying relationship, and resolving conflicts with solving circles. Coaching requires couples to demonstrate and practice strategies provided during the sessions. The reality therapist may employ techniques such as the making money and the photo album activities and the WDEP system (Wubbolding, 2000).

During the first session, the reality therapist begins with identifying the couple's goals by using the WDEP system (Wubbolding, 2000). The session begins with the counselor asking, "What do you *want* ("W") to gain from premarital counseling?" After identifying the wants/ goals, the counselor can begin to assess the relationship. Kimberly and Desmond identified improving their communication as a goal for counseling. Continuing with the WDEP system, the counselor focused on the "D" by asking the couple to identify what they were currently *doing* to improve their communication (their total behaviors of acting, thinking, feeling, and physiology). Once the couple has described their current communication patterns, the reality therapist may ask, "Are your behaviors strengthening or weakening your relationship?" Self-evaluation is a critical step in reality therapy (Glasser, 1998; Wubbolding, 2000, 2011). Once behaviors have been identified, each person must evaluate ("E") his or her behavior to determine if the behavior is contributing to the goal. The final step is to assist the couple in developing a plan ("P") to obtain their desired goal (Wubbolding, 2000, 2011).

Techniques

Kimberly and Desmond's primary goal for counseling was to improve their communication. In addition to observing the couple's interaction during the session, a reality therapist can employ a variety of techniques to assess the couple's communication patterns.

The Picture Album   One of these techniques includes the assignment of homework. The therapist asked Kimberly and Desmond to create a collage of pictures or an "album" representing their ideal marriage. The couple should create individual albums and then share them with each other before the next counseling session. The goal was to find the commonality in the couple's relationship (Wubbolding, 1988) because good relationships are based on common pictures.

Because premarital counseling involves education as well as counseling, the reality therapist will also teach the couple choice theory principles. The "picture album" activity will allow the therapist to introduce the concepts of the quality world. An individual's quality world consists of a group of specific pictures that often illustrate whom he or she wants to be with, things he or she wants to own or experience, and beliefs that direct his or her behavior (Glasser, 1998). Glasser and Glasser (2007) state:

> The pictures in our quality world are the actual motivation for all our behavior. While needs are the genetic source, it is these very specific pictures of the way we want to live our lives that cause us to do whatever we do from birth to death. (p. 53)

As the couple explores the pictures in their quality world, they can begin to identify and define what they want in their relationship. This activity can be very enlightening for couples because they have an opportunity to articulate their wants. For some couples, this may be the first time the individual shared his or her wants with the partner. The more detailed the picture is, the greater the possibility that it will be satisfied (Wubbolding, 1998). Couples should share and update the pictures in their albums periodically during the marriage because pictures change, are revised, and/or replaced as the marriage progresses (Wubbolding, 2000).

When the couple returned to the next session, they were asked about the homework assignment. Desmond stated, "It was fine, but Kimberly did not like some of my pictures. She started criticizing me for not including her more." Desmond had pictures of playing and watching sports. He also had pictures of playing tennis with Kimberly.

Kimberly said,

I like sports. I like playing tennis but I don't want to spend all of my weekend playing or watching sports on television. I feel like he does not care about what I like to do. I think he needs to grow up.

The reality therapist explained that we all have pictures in our quality world and good marriages are built on common or shared pictures:

You both had pictures of playing tennis with each other in your album. Tennis is an activity that both of you enjoy together, so you do have a common picture. Your albums do have different pictures and, as a couple, you will have to learn how to negotiate your different pictures.

The therapist asked, "Desmond, how do you feel about Kimberly's comments?"

Desmond replied, "I don't like what she said. I hate when she criticizes me because I do care about her. I am marrying her."

Caring and Deadly Habits   The therapist then introduced the concept of the caring and destructive behaviors (Glasser & Glasser, 2000). After learning these principles, the couple can develop positive ways to communicate their wants.

Deadly habits or disconnecting behaviors occur when a person chooses to coerce, force, blame, criticize, nag, punish, or bribe and/or threaten another person (Glasser & Glasser, 2000, 2007). When couples employ these behaviors, they may harm or even destroy their relationships. The use of the deadly or disconnecting habits will move couples farther apart and intensify the problem (Glasser & Glasser, 2007). Using the deadly habits is a result of the principle of external control. Contrary to external control, choice theory proposes that an individual can only control his or her own behavior. One cannot control another person's behavior (Glasser, 1998; Glasser & Glasser, 2000, 2007). Therefore, nagging, criticizing, or blaming the partner will not change his or her behavior; in fact, it may result in damaging the relationship.

To create and maintain closeness in relationships, couples must avoid deadly habits and communicate using the caring habits. The caring habits or connecting behaviors are supporting, listening, trusting, accepting, and trusting (Glasser & Glasser, 2000, 2007). The goal is to teach couples to communicate in a way that strengthens the connection in their relationship. Glasser states, "As simple as this seems, it is almost impossible to do unless you recognize how much harm you are doing to your marriage with the deadly habits" (Glasser & Glasser, 2007, p. 35).

Making Money   To help couples recognize when they are communicating using a deadly or caring habit, a technique called *making money* can be used. This technique uses banking principles to represent caring and deadly habits. When a couple is communicating using caring habits, they are making *deposits* into their relationship account or "making money." When the couple uses deadly habits, they are making *withdrawals* from their relationship account. Too many withdrawals can result in an overdrawn account, which can lead to a closed account. Couples are encouraged to make deposits daily and avoid making withdrawals. The deposits and withdrawals can be recorded on a spreadsheet or the *relationship balance sheet*.

The guidelines to implementing this technique are the following:

- If someone says a caring comment, the receiver would say, "You are making money" and follow the comment with a sign of affection (i.e., kiss, back rub, etc.).
- If someone says a deadly comment, the receiver would respond by saying "insufficient funds" and walk away to avoid an argument or responding with a deadly comment.
- Once a person hears "insufficient funds," the person needs to self-evaluate the comment and replace it with a caring comment.
- If the speaker chooses not to replace the deadly comment with a caring one, then the speaker should walk away without making any further comments.

- The speaker cannot explain or spend time justifying the comment. If the receiver perceived the comment as deadly, then the speaker has to accept it.
- A monetary value can be given to the deadly and caring comments—for example, $10 for each caring or deadly comment. When an individual communicates using a caring comment, a $10 deposit is made into the relationship account. When a deadly comment is spoken, a withdrawal of $10 is made from the relationship account.

At the end of each day or a predetermined time period, the couple can review their account to determine if they *made money* or have *insufficient funds*. This technique can serve as a trigger for the couple to self-evaluate their communication style.

When Kimberly and Desmond returned for the third session, they were extremely anxious to share their albums and the relationship balance sheet. Kimberly stated, "We talked about kids, household chores, cars, vacations, furniture and many other things. We talked about topics we had never discussed before."

Desmond said, "When we shared our albums, we found many common pictures. It was great to see how much we have in common, but there were some differences." When the therapist asked how the differences were handled, Kimberly said, "I did not criticize him. I used caring habits to communicate. I made a lot of money this week." Desmond agreed and smiled. Desmond stated, "This activity has made me recognize when I am using deadly habits and to evaluate my words. I am listening more and using more caring language. I love to hear, 'You are making money, Baby.'"

WDEP In addition to teaching the couple about the quality world and caring habits, the reality therapist will teach a couple how to make good choices in their marriage. The WDEP system is a pedagogical tool that can teach individuals how to make better choices, have better control over their lives, and get their needs met (Wubbolding, 2000, 2011). The WDEP system is introduced to the couple during the first session. Having the couple create their picture album can be considered a first step in determining what the couple wants in their marriage. Wubbolding suggests that, through the use of skillful questioning, the reality therapist will help the couple determine what they want from their marriage from their spouse, in-laws, etc. (2000). The couple can clarify desired characteristics of their marriage, which can lead to effective planning (Wubbolding, 2000).

Because Kimberly and Desmond wanted to improve the communication in their relationship, the therapist might ask them, "What, specifically, do you want to improve?" The more detailed the want is, the greater the likelihood of satisfying it is (Wubbolding, 2009).

Once the couple has identified what they want from the marriage, the therapist can ask them the following question: "What are you doing to improve your communication?" The reality therapist helps the couple describe current behaviors (Wubbolding, 2000). Since reality therapy is built on the principle of internal locus of control, each person will have to answer the question by focusing on his or her behavior (Glasser & Glasser, 2000; Wubbolding, 2000, 2009, 2011). Each person lists his or her behaviors, positive and negative. The therapist may also ask couples, "Are you headed in the direction you want?" Wubbolding wrote, "Often people don't see their direction until they describe the specifics of their 'doing' or behaviors" (2009, p. 19). Once behaviors have been identified, the therapist can assist the couple with self-evaluation.

It is not enough to have the couple identify their behaviors. A reality therapist wants the individuals to make judgments about their behaviors (Wubbolding, 2000). Self-evaluation is the core of reality therapy (Glasser, 1998). "The reason that this procedure is so important is that many people repeat behaviors that are not helpful and sometimes even harmful" (Wubbolding, 2000, p. 111). A reality therapist would ask, "How is that working for you?" or "Is that helping you get what you want?" When Kimberly was completing the *making money* exercise, she described her behaviors as making critical comments to Desmond, as well as giving him the silent treatment. The therapist can ask Kimberly, "Is that behavior helping or hurting your relationship with Desmond?" If the behavior is helping the relationship, then it should continue. If the behavior is hurting the relationship, the therapist will help the client replace it with a more effective behavior. Wubbolding states, "No one changes behavior unless a judgment is first made that current behaviors are not helpful" (2000, p. 111).

To help couples make meaningful change in their lives, the reality therapist will help the couple develop a plan to obtain what they want. Kimberly and Desmond wanted to enhance their communication skills. They described the behaviors that they were currently doing and evaluated their effectiveness. The reality therapist would ask the couple, "What is your plan for change?" The therapist believes the characteristics of a good plan are $SAMI^2C^3$: simple, attainable, measurable, immediately involve the therapist (if needed), controlled by client, commitment, and consistent/repeatable (Wubbolding, 2000, 2009, 2011).

The best plan is one developed by the client (Wubbolding, 2000). Kimberly and Desmond decided that a good plan for them was to keep using the making money technique. It was easy for them to do. They could start tracking their caring and deadly comments immediately. The balance sheet helped them measure their progress. They were both responsible for recording their own behavior. Finally, Kimberly and Desmond were committed to doing it every day. Desmond suggested having a meeting to review the account or "balance the checkbook" once a week. Kimberly agreed with the plan and decided to post it on the refrigerator.

The fourth session began by asking the couple to share what they had learned from the counseling sessions. Kimberly stated, "I know Des better. I know what he wants and I have learned how to talk in a supportive manner." Desmond said, "We took our individual albums and created a marriage album with our common pictures. I feel like our foundation is stronger than before." The couple said they are continuously evaluating their behavior. Kimberly said, "I am always saying, 'Is my behavior helping or hurting the situation?'" Desmond stated, "We are stronger than ever. I know our marriage will be better because we have better communication and the tools to make good decisions." The goal of the final session is to summarize the information that has been presented and to address the couple's unanswered questions.

DISCUSSION

Kimberly and Desmond attended premarital counseling as requested by their minister. Their goal was to enhance the communication in their relationship but they did not identify any other problems. The reality therapist used the sessions to teach the couple about choice theory and reality therapy principles. The couple identified their wants by developing a marriage picture album. They learned to communicate by using the caring habits and avoiding the deadly habits. The making-money technique was implemented to help the couple recognize and evaluate their behavior. To help the couple satisfy their needs, the WDEP system was taught. Like most engaged couples, Kimberly and Desmond left premarital counseling with a clearer picture of marriage and each other.

SUMMARY

Premarital counseling is unique because the couples are not coming to counseling to address issues. Frequently, the couples are so excited about getting married that they fail to discuss key issues. Many couples build their relationship on external control or the premise that they can change or control their partner. Teaching choice theory and reality therapy to couples can provide the tools to build a healthy marriage. "Choice theory does not guarantee a wonderful marriage; it guarantees a way to deal with the problems that will come up in the best marriages" (Glasser, 1998, p. 175).

When engaged couples begin their marriage understanding the "wants" or pictures in each person's quality world, they have a better understanding of each individual's expectations and aspirations. Couples now understand that each partner's behavior is related to satisfying a need. They have several tools in their marriage tool belt to build their marriage:

> To get along better than we do now with another person, we need to try to learn what is in that person's quality world and then try to support it. Doing so will bring us closer to that person than anything else we can do. (Glasser, 1998, p. 51)

Choice theory and reality therapy can educate couples on positive ways to communicate. By learning the difference between deadly and caring habits, couples can avoid disconnecting in their relationships. When couples use the WDEP system, they can identify their wants, evaluate their behaviors, and develop effective plans. Reality therapy provides a "practical structure and delivery system" (Wubbolding, 2000, p. 82).

Premarital counseling is different from traditional marriage counseling. Premarital counseling is proactive and not reactive. The therapist does not want to search for problems or attempt to create problems to fix. Premarital counseling is designed to provide engaged couples with techniques to create a great marriage. The strategies used in premarital counseling can be implemented with married couples. Choice theory and reality therapy can help couples build strong relationships that can lead to happy marriages.

REFERENCES

Glasser, W. (1998). *Choice theory: A new psychology of personal freedom.* New York, NY: HarperCollins.

Glasser, W., & Glasser, C. (2000). *Getting together and staying together.* New York, NY: HarperCollins.

Glasser, W., & Glasser, C. (2007). *Eight lessons for a happier marriage.* New York, NY: HarperCollins.

Markman, H., Rhoades, G., Stanley, S., Ragan, E., & Whitton, S. (June, 2010). The premarital communication roots of marital distress and divorce: The first five years of marriage. *Journal of Family Psychology, 24*(3), 289–298.

Moss, D. C. (1988). Divorce survey. *American Bar Association Journal, 72*(2), 30.

National Center for Health Statistics (2010). Births, marriages, divorces, and deaths: Provisional data for 2009. *National Vital Statistics Reports, 58*(25). Hyattsville, MD.

Owen, J., Rhoades, G., Stanley, S., & Markman, H. (February, 2011). The role of leaders' working alliance in premarital education. *Journal of Family Psychology, 25*(1), 49–57.

Wubbolding, R. E. (1988). *Using reality therapy.* New York, NY: Harper Perennial.

Wubbolding, R. E. (2000). *Reality therapy for the 21st century.* Philadelphia, PA: Brunner-Routledge.

Wubbolding, R. E. (2009). *Reality therapy training manual* (15th ed.). Cincinnati, OH: Center for Reality Therapy.

Wubbolding, R. E. (2011). *Reality therapy: Theories of psychotherapy series.* Washington, DC: American Psychological Association.

IV
Conclusion

17

COMMENTARY FROM THE EDITORS

Patricia A. Robey, Robert E. Wubbolding, and Jon Carlson

INTRODUCTION

In Chapter 8 of *Choice Theory* (1998), William Glasser discussed some of the problems inherent in love and marriage. He noted that we are drawn to relationships because it feels good; we hope we have found a partner who will accept us for who we are as well as who we are trying to be. Glasser stated, "It is this willingness, even eagerness, to share your hopes and fears that defines love. As long as you can do so, you have a very good chance of staying in love" (p. 164).

As we have seen in the case studies presented in this book, problems occur when couples no longer accept their partners for who they are. Instead, they try to change their partners by using coercive and controlling behaviors that disconnect them even further:

> External control is very simple. In a relationship it is a belief that what we choose to do is right and what the other person does is wrong. ... That external control attitude, *I know what's right for you*, is what people driven by power use when they are in an unhappy relationship. One or both may use it but even if only one uses it consistently it will eventually destroy that relationship. (Glasser, 1998, pp. 20–21)

Learning choice theory concepts and applying them in their relationships can help couples reignite the spark that initially drew them together.

In this chapter, the editors provide their final commentary on the preceding chapters and on the concepts of choice theory and reality therapy. The editors hope that the information provided in this book will be useful to you as you work with challenging couples. We encourage you to use choice theory and the reality therapy process to understand yourselves as well as your clients. When you feel challenged by your clients, you can ask yourself

What is it that *I* want to see happen in the session right now? What am I doing [actions, thinking, feelings, and physiology] to get what I want? What is working for me? What is not working? What can I do differently to get what I want right now?

COMMENTARY FROM PATRICIA ROBEY

Theory explains human behavior and motivation; therapy is the process we use to put the theory into action in counseling. The value of having a theory to support your work in counseling is that theory provides a framework for understanding your clients before they even come to see you. For example, from a choice theory viewpoint, I know that clients have five basic needs, which they attempt to meet through very specific people, things, values, and beliefs. People choose behaviors in an attempt to get the things or relationships that they want, which satisfies one or more of their basic needs. The results of their behavior lie on a continuum from effective to ineffective.

Reality therapy is the application of choice theory concepts in counseling. Choice theory explains that the main source of human unhappiness is difficulty in relationships. Therefore, when clients come to counseling, I know that they are having problems with relationships. Relationship problems arise when we try to control others in an effort to get them to do or be something different from what they are. The goal of therapy, then, is to help clients reconnect with others. As demonstrated in the case studies presented in this book, counselors focus on the present and avoid discussion of symptoms and complaints. Reality therapy is strength based and encourages clients to change their own behaviors rather than put energy into what they cannot change: the behavior of others.

It Sounds Simple, but It Is Not Easy

I remember my mentor, Bob Wubbolding, saying that reality therapy "sounds simple but it's not easy." He was correct! Clients are often invested in being *right* and want to maintain the belief that it is other people who are causing the problems in their relationships. After all, admitting that we are part of our own problem is humbling—a direct hit to our need for internal power and feelings of competence! In spite of our best efforts as counselors, our clients will behave in counseling to get their own needs met. This may mean that they try to triangulate counselors into their relationships so that they feel a sense of love and belonging with their counselors; they argue their points in order to feel powerful; they skip appointments or renege on homework plans to gain a sense of freedom; they attempt to distract from the hard work in sessions so that they can have some fun; they walk out of sessions when they feel so angry that their very survival may feel threatened.

When you read the chapters and the case studies, you may have thought this approach sounded too simplistic. The challenge in writing a book like this is that the chapters provide only an introduction to the topics and a brief overview or summary of this process in action. I hope that you will find inspiration in the chapters and will take the time to further explore the ideas of choice theory and reality therapy and the application of these ideas in other areas of interest to you.

Evolution of Ideas and Relevance in 2011

The evolution of these ideas began in 1954, when Glasser worked with veterans who were diagnosed with schizophrenia, and it continued to develop as Glasser worked at the Ventura School for Delinquent Girls. In 1965, Glasser wrote *Reality Therapy* and began work in Watts in the Los Angeles Unified Public School District. *Schools Without Failure* was published in 1969 and Glasser's transformative work in education continued with the development of the Educator Training Center, which focused on helping schools create systemic change. By 1974, Glasser's books had been translated and sold all over the world. Glasser would continue to develop, apply, and write about his ideas in books like *Take Effective Control of Your Life* (1984; reprinted as *Control Theory* in 1985), *Control Theory in the Classroom* (1986), *The Quality School* (1990), *The Quality School Teacher* (1993), and *Choice Theory* (1998).

In the twenty-first century, Glasser's application of choice theory and reality therapy continued to evolve as he addressed systemic issues in schools with *Every Student Can Succeed* (2000b). Glasser also continued to be an advocate for mental health, writing books on counseling and psychiatry—*Warning: Psychiatry Can Be Hazardous to Your Mental Health* (2003) and *Defining Mental Health as a Public Health Problem* (2005). The reality therapy process is especially useful today as managed care influences the number of sessions available for clients. Due to its focus on strengths, making plans for action, and empowering clients by teaching them problem-solving skills, reality therapy can be successfully used in a brief time. It is applied in many contexts, including education, corrections, physical and mental health management, business, counseling, and personal development. These ideas continue to be taught worldwide and are appropriate for use across cultures.

Choice Theory and Reality Therapy as a Couple and Family Approach

As demonstrated in this book, reality therapy and choice theory are not only useful for individual counseling, but also can be effective in work with couples, families, groups, and systems. Because relationship issues are considered to be the primary problems affecting human beings, it is natural to bring both, or all, individuals involved in the relationship system into the counseling setting. Teaching couples and families to understand their roles in the problem relationship and helping them to define mutual wants and goals for change are useful in breaking the cycle of ineffective behavior that permeates problematic interactions.

Many choice theory and reality therapy publications have focused on couple and family issues and/or proactive approaches to relationship interactions. Recent books by Glasser include *What Is This Thing Called Love?* (Glasser & Glasser, 2000b), *Getting Together and Staying Together* (Glasser & Glasser, 2000a), *Unhappy Teenagers* (Glasser, 2000c), and *Eight Lessons for a Happier Marriage* (Glasser & Glasser, 2007). Olver's book, *Secrets of Happy Couples: Loving Yourself, Your Partner and Your Life* (2011), discussed the application of choice theory and reality therapy in building stronger relationships. Included in Olver's book is information from relationship experts that supports the use of choice theory and reality therapy in couples counseling. This support comes from the ideas of Gary Chapman, John Gray, Harville Hendrix, and others. Since 2010, nine articles on couple or family therapy have been published in the *International Journal of Choice Theory and Reality Therapy*, and a 2011 PsycInfo search of keywords "William Glasser" resulted in 647 results, with 106 results when the keyword "family" was added.

COMMENTARY FROM ROBERT WUBBOLDING

The authors of the chapters in this book discuss a wide range of choice theory/reality therapy applications for strengthening relationships. These illustrations of the practice of reality therapy are intended to be practical and useful for therapists, counselors, and anyone desiring to enrich his or her life.

Topics such as multicultural counseling, recovery after infidelity, substance abuse, caregiving, marriage preparation, counseling celebrities, and, in fact, all chapters present useful insights that can be generalized beyond the specific relationship discussed by the respective authors. Skills and techniques presented offer expanded benefits that the reader can employ in a variety of settings, including the therapy office and the home. These practical ideas can be used by you, the reader, in your own life as well as being integrated into other counseling modalities: Adlerian, cognitive, and multimodal, as well as the current diverse assortment of brief therapies.

In learning from the narratives included in this book, readers are advised to understand that the authors have summarized a process too lengthy to be described in all its detail. Of necessity, much dialogue and many details are omitted, so the development of the counselor/client relationship and the progress of the couple might appear to be more unadorned and less complicated than they really are.

Ethics

Similarly, in view of the fact that the cases are presented in the context of reality therapy and its underlying principles of choice theory, the authors have not included issues focusing on ethical principles. All professional organizations require that counselors and therapists work within the boundaries of both ethics and the law. Because such counselor/client discussions are excluded from the text does not imply that reality therapists ignore standard practice or the standard of care. The purpose of this book is not to present information about the requirements of counselors and therapists to practice within ethical standards. All counselors and therapists, including reality therapists, deal openly with issues such as informed consent, professional disclosure, boundaries, multiple relationships, and confidentiality and its limits, as well as other ethical issues described in the ethical codes of psychologists, counselors, and social workers.

Take What Is Useful …

The cases presented by each counselor described throughout this book illustrate the work of each particular counselor. The expertise, experience, knowledge, and worldview of the helper always guide the application of reality therapy. Therefore, readers will find therapists' comments and methodologies with which they agree and disagree. You are encouraged to take what is useful and leave aside what is not useful. If readers conclude that they have a better idea, a more effective tool, or a more relevant intervention for particular cases, the book has been a success. Both the ideas and the specific interventions presented in this book are intended to be useful to the reader. But also, the editors have nourished a hopeful anticipation that the narratives trigger the reader's thinking and thus assist you in expanding your repertoire of ideas, skills, techniques, and positive outcomes.

This book will also be a success if you, the reader, come to two beliefs. One is that reality therapy is an efficacious methodology. It works. Second, the editors believe, on the deepest level, that human behavior is changeable (i.e., we are not locked into positive and effective choices or imprisoned by destructive or even mildly harmful choices). Though some philosophers throughout history have subscribed to the naïve belief that human nature is perfectible, these editors accept the fact that, each day, human beings make choices and develop habits.

But at times people act against their long-standing and seemingly permanent patterns of behavior. The popular media flood the news with information about individuals who have a long history of success and ethical lifestyles who, as it were, "fall off the wagon" and disgrace themselves. On the other hand, without publicity, many individuals who have been addicted, committed heinous crimes, or led mediocre lives often alter their life paths by making unexpected and momentous changes in their behavior. They choose heroic, altruistic, or, at least, more effective lifestyles.

When counseling individuals with mild or severe problems that have originated from within themselves or clients who have severe limitations or pain thrust upon them from their environment, reality therapists firmly believe that a better life is possible. This is not the same as believing that change is easy or that the principles of reality therapy inevitably and absolutely lead to self-actualization or earthly happiness. It is standard practice in the helping profession to speak of "improvement"—not cure. Because human beings choose individual, specific actions, it is never appropriate to guarantee results.

If You Do Not Use the Ideas, They Will Not Work

There is only one guarantee that these authors present to the readers: If you do not use the ideas, they will not work. Therefore, we encourage you to implement at least one idea immediately either in your own personal lives or in your professional lives.

To the students of counseling and psychotherapy, we encourage you to keep your idealism. If you believe you can change the world, do everything you can to keep that ideal before you. If you help one client or one couple to make better choices, you not only impact them but also impact their progeny. When you teach couples through your counseling that they can control their lives more effectively, you have played a part in their journey and in their effort to achieve their destinies. Moreover, your influence ultimately cascades down through history into the future. Providing help for clients by listening, empathizing, helping them define what they want, evaluating their behavior, and planning for the future, you impact the people in your immediate care and also, indirectly, the people they live with or encounter in other ways. Finally, in choosing a career in the helping professions, you have selected one of the noblest endeavors conceived by the human mind.

COMMENTARY FROM JON CARLSON

Creating satisfying relationships is at the heart of choice theory and reality therapy. The important concepts have stood the test of time as well as empirical scrutiny. The core ingredients needed to be an effective person turn out to be the same ones needed to be an effective partner. The core ingredients that Dr. Glasser has isolated seem to be as follows:

> *Choice.* In life as well as marriage we make choices that lead to happiness or misery. No one else can make these choices. Unhappy partners give up this power and claim they are trapped or unable to change their situations. Learning to accept that we make important choices throughout the day, we choose to fight, keep quiet, be reasonable, etc. Partners can learn that in healthy marriages, both partners choose to keep their voices and collaboratively work through areas of difference.
>
> *Self-responsibility.* Happy partners concentrate on changing themselves rather than their partner's behaviors. They have learned that acceptance of their partner's strengths *and* liabilities will go far toward creating a long-term satisfying marriage. The focus is on self-control rather than external control or trying to control the partner. All of our behaviors are chosen, so we can stop and not participate in unproductive, unsatisfying dialogue.
>
> *Creating happiness.* When you do not like something, do something different. Do not blame your partner; instead, change the focus and understand how you participate in creating

the misery. We are only as trapped as we choose to be. Our thoughts often seem like facts and not just thoughts. Thoughts can be changed. Once you become aware of the personal power to change thoughts, you can change misery to satisfaction. We can create happiness where misery or mediocrity once existed.

Mental health. Mentally healthy people strive to get along with all the significant people in their life. Relationship satisfaction is paramount to the ability to get along with all kinds of people, from relatives to friends to strangers. By focusing on having positive relationships with others, we notice that the conflict in other areas of our life tend to decrease significantly. Mental health and happiness are rooted in the ability to be connected in positive ways to others.

The relationship. Too often in partnerships, couples argue over who is going to get his or her way. Is it best for her? Is it best for him? Reality therapy stresses the importance of the relationship and teaches couples to ask the question, "What is best for us? What does our quality world look like?" As couples clearly discuss "who we are," this becomes their compelling future. A compelling future allows the couple to have a clear understanding as to who they are and where they are headed with their life.

Seven deadly habits. Couples must avoid the seven deadly habits if they want a satisfying partnership. The Glassers identify the negative schemes as criticizing, blaming, complaining, threatening, nagging, punishing, and bribing or rewarding to control. They can be replaced with supporting, encouraging, listening, accepting, respecting, and negotiating differences. No couple can do this 100% of the time, but they should strive toward improving the balance of positive over negative habits in their exchanges.

Creativity and doing the unexpected. All people like novelty, which is the "spice of life." If you are not sure of what to do or say in a situation, then do something different. It is easy to get into a relationship "rut" of doing the same thing over and over even though it does not lead to happiness or satisfaction.

Love. Love is present in most relationships, although it is not clearly defined. A reality therapist stresses that *love is the wish to make another happy* and keeps the focus on being a good partner and not demanding that the partner be a good one. Love has more to do about others' happiness rather than our own. It is through giving love that we receive it.

CONCLUSION

The concepts of choice theory and the practice of reality therapy are taught and are practiced on every continent except Antarctica. Wubbolding (2000, 2011) described the relationship between choice theory and reality therapy using the metaphor of track and train. Choice theory provides the track to guide the reality therapy train as the delivery system used when working with clients. Reality therapy has been used by thousands of people since its development in 1965, and it continues to be a pragmatic approach to counseling today.

The William Glasser Institute offers a variety of training opportunities, including an 18-month process that leads to certification in choice theory and reality therapy. For further information on training or applications of choice theory and reality therapy, we encourage you to visit the William Glasser Institute at www.wglasser.com.

REFERENCES

Glasser, W. (1965). *Reality therapy.* New York, NY: Harper & Row.

Glasser, W. (1969). *Schools without failure.* New York, NY: Harper & Row.

Glasser, W. (1984). *Take effective control of your life.* New York, NY: Harper & Row.

Glasser, W. (1985). *Control theory.* New York, NY: Harper & Row.

Glasser, W. (1986). *Control theory in the classroom.* New York, NY: Harper & Row.

Glasser, W. (1990). *The quality school.* New York, NY: HarperCollins.

Glasser, W. (1993). *The quality school teacher.* New York, NY: HarperCollins.

Glasser, W. (1998). *Choice theory.* New York, NY: HarperCollins

Glasser, W. (2000a). *Reality therapy in action.* New York, NY: HarperCollins.

Glasser, W. (2000b). *Every student can succeed.* Chula Vista, CA: Black Forest Press.

Glasser, W. (2000c). *Unhappy teenagers.* New York, NY: HarperCollins.

Glasser, W. (2003). *Warning: Psychiatry can be hazardous to your mental health.* New York, NY: HarperCollins.

Glasser, W. (2005). *Defining mental health as a public health problem.* Chatsworth, CA: The William Glasser Institute.

Glasser W., & Glasser, C. (2000a). *Getting together and staying together.* New York, NY: HarperCollins.

Glasser W., & Glasser, C. (2000b). *What is this thing called love?* New York, NY: HarperCollins.

Glasser W., & Glasser, C. (2007). *Eight lessons for a happier marriage.* New York, NY: Harper Collins.

Olver, K. (2011). *Secrets of happy couples: Loving yourself, your partner and your life.* Chicago, IL: Inside Out Press.

Wubbolding, R. (2000). *Reality therapy for the 21st century.* New York, NY: Routledge.

Wubbolding, R. (2011). *Reality therapy: Theories of psychotherapy series.* Washington, DC: American Psychological Association.

GLOSSARY

Basic needs: Genetic instructions that provide the underlying motivation for behavior. They include survival, love and belonging, power, freedom, and fun.

Caring habits: Seven behaviors that are likely to improve relationships: listening, supporting, encouraging, respecting, trusting, accepting, and always negotiating disagreements.

Choice theory: A theory that describes human behavior and motivation as originating within a person, not as a result of external stimuli from the environment or past experience. Choice theory is the foundation for the practice of Reality Therapy.

Deadly habits: Seven behaviors that are likely to damage relationships: criticizing, blaming, complaining, nagging, threatening, punishing, and bribing or rewarding to control.

External control psychology: The idea that people and things outside ourselves can directly determine our behavior and that, by manipulating the environment, we can control another's behavior, whether that person wants us to or not.

Perception: Information received from the external/real world as we see it; this may differ from how others view the same information.

Perceptual filters: The lenses through which we recognize and place value on the information we perceive from the external/real world, including sensory, knowledge, and value filters.

Quality world: The mental pictures we hold of the people, places, things, behaviors, and values that satisfy one or more of the five basic genetic needs; the proximate source of motivation.

Real world: The external world; people, situations, and things that actually exist.

Reality therapy: The therapeutic process developed by William Glasser.

Solving circle: A metaphorical place in which a couple, family, or group agrees to use caring habits when discussing issues or problems they are experiencing in their relationship.

Structured reality therapy: A six-step process used to help couples shift their focus from problem thinking and blaming to making a commitment for relationship change by choosing more connecting behaviors.

Total behavior: The actions, thinking, feeling, and physiology that occur simultaneously; total behavior is chosen and generated in an effort to get what we want (quality world pictures), which satisfies one or more of the basic needs.

WDEP: A system for delivering reality therapy developed by Robert Wubbolding based on choice theory; it provides a framework designed to help therapists work with clients by exploring clients' **w**ants (quality world pictures) and **d**oing (total behavior), helping clients **e**valuate the effectiveness of their behavior, and then making a **p**lan of action for more effective behavior.

INDEX